SCIENCE COMMUNICATION

An Introduction

World Scientific Series on Science Communication

Editor-in-Chief
Hans Peter Peters
Forschungszentrum Jülich, Germany
h.p.peters@fz-juelich.de

Aims and Scope

The books published in this series deal with the public communication of science, i.e. with communication of and about science involving non-scientists and taking place in the public sphere. Possible topics include public communication and discourses about scientific knowledge, scientific projects or research fields, and communication about science as a social system and its interdependencies with the larger society.

This book series is open to analyses of all forms of public communication and interaction: journalism, public relations of science, blogs, social media, video-sharing platforms, science museums, public events, engagement activities, public deliberation and participation, citizen science and other collaborations between scientists and citizens, for example. Books may focus on content and processes of messages and discourses, on actors and their strategies, on channel and arena characteristics, on the reception, effects and use of public expertise, on public controversies, on inclusion of citizens in public discourses, and on issues of quality, ethics, and trust.

Typically, authors/editors will come from the academic field and have an academic audience in mind. But some of the books may also be relevant for communication professionals, scientists (as communicators), science managers and knowledge users, for example. Books may be based on specific research projects, deal with a relevant subjects by means of review of existing studies and theoretical discussion, or publish contributions of a relevant conference (proceedings).

Published

World Scientific Series on Science Communication — Volume 1

SCIENCE COMMUNICATION

An Introduction

Editors

Frans van Dam (Utrecht University, The Netherlands)

Liesbeth de Bakker (Utrecht University, The Netherlands)

Anne M Dijkstra (University of Twente, The Netherlands)

Eric A Jensen (Institute for Methods Innovation, UK)

W\|Ͼ World Scientific

N JERSEY · LONDON · SINGAPORE · BEIJING · SHANGHAI · HONG KONG · TAIPEI · CHENNAI · TOKYO

Published by

World Scientific Publishing Co. Pte. Ltd.

5 Toh Tuck Link, Singapore 596224

USA office: 27 Warren Street, Suite 401-402, Hackensack, NJ 07601

UK office: 57 Shelton Street, Covent Garden, London WC2H 9HE

British Library Cataloguing-in-Publication Data
A catalogue record for this book is available from the British Library.

World Scientific Series on Science Communication — Vol. 1
SCIENCE COMMUNICATION
An Introduction

ISBN 978-981-120-987-1 (hardcover)
ISBN 978-981-120-988-8 (ebook for institutions)
ISBN 978-981-120-989-5 (ebook for individuals)

For any available supplementary material, please visit
https://www.worldscientific.com/worldscibooks/10.1142/11541#t=suppl

Contents

Foreword

Efforts to communicate science to the general public may be traced back to the early days of science in the 16th, 17th and 18th centuries. Several countries around the globe have outstanding and even heroic examples of talented scientists who wrote articles and books for lay audiences, gave public lectures or presented popular science shows. Their main objective was to enlighten people about different science topics or issues, or even simply to entertain them. Since the 1950s, a modern era of science communication has emerged around the world. This modern era is characterized by a definite and intentioned movement toward incorporating science into the general culture of a population. In the pioneering years of this modern era, science communicators were very much like their colleagues in earlier centuries enthusiastic volunteers aiming to enthuse the general public about science, but frequently unappreciated.

Since those early years, science communication has matured into a rich and complex field which is evolving within a fast-changing and expanding communication landscape. At the same time, science itself is increasingly influenced by political, commercial, and institutional interests. Scientific knowledge has become an essential ingredient of solutions for many of the challenges modern societies face today such as climate change, threatened biodiversity, pollution, sustainable energy, public health, and decreasing natural resources. However, scientific knowledge alone is not enough. Such solutions require public policies, which, in turn, require the participation of a society composed of individuals who are capable of making informed decisions concerning issues rooted in their natural and social environment. The task of facilitating such civic dialogue is beyond the scope of the formal educational system. Therefore, the communication of science through various media, using different strategies aimed at reaching specific sectors of society, has become an essential activity in modern knowledge societies.

As a result, science communication has evolved from being a voluntary, empirical and part-time endeavor for those who had other occupations such as research, teaching, or journalism to also being a full-time professional activity which requires specialized knowledge and skills. Due to the complexity of the field, a range of professional profiles, backgrounds, and expertise is needed. As science communication evolves and matures as a field of professional practice, scholarship and academic training, an increasing number of educational programs for science communicators, ranging from short courses to PhD programs, has been on offer at research institutions and universities around the world. The spectrum is wide and varied, with different entry and exit requirements, curricula, duration, and certifications. Some specialize in certain scientific issues or topics while others specialize in specific types of media. Some of these have a practical focus, while others aim to develop science communication scholarship and research expertise.

All these programs require up-to-date and comprehensive reading materials. This book, *Science Communication: An Introduction*, responds to this need. It provides a comprehensive and current overview of how this field has evolved, as well as theoretical basics and strategies that science communication practitioners and research students must master. Adding to its value and significance, this handbook helps to bridge theory and practice in science communication by providing a series of real-world case studies.

Since the relationship between science and society is influenced by historical, bio-geographical, cultural, and socio-political factors, there are differences in how public communication of science has developed in different countries and how public communication of science is practiced today. The relevance of this handbook is therefore enhanced by the inclusion of examples and views from several countries, in particular by including some perspectives from the developing world. Although it focuses mainly on European culture, it contains ample elements and references that will help readers and science communication students appreciate the different characteristics and priorities of public communication of science in other world regions.

For newcomers who are interested in science communication research and practice, this book follows the logistics of showing the landscape of the domain step-by-step, unfolding the whole story of how science relates to the public and social development. Each chapter includes rich literature references which show the editors' and authors' expertise in the field, with extensive arguments and different points of view providing the readers with inspiring elements which help them arrive at their own conclusions. 'Authoritarian approaches' as well as excessive analysis are avoided. But, stimulating open questions are provided for further thought.

We, the reviewers of this book, come from countries with different social and cultural contexts. However, the cases of science communication research and practice presented in this book are adaptable to our different cultures. Centering on key theoretical issues, the book offers evidence-based cases presenting science communication as a social and cultural phenomenon of different stakeholders as well as the public's expectations related to science and technology.

We welcome this handbook that will be useful for science communication students, researchers, and practitioners, and function as a meaningful introduction for science policy-makers.

Elaine Reynoso-Haynes
Director of Training and Research in Public Communication of Science
at the Dirección General de Divulgación de la Ciencia, Universidad Nacional
Autónoma de México

Marina Joubert
Senior Science Communication Researcher and Lecturer at
the Centre for Research on Evaluation, Science and Technology
at Stellenbosch University, South Africa

Yin Lin
Associate Researcher and Deputy Director of the
Division of Science Popularization Policy Research at
the China Research Institute for Science Popularization

List of Contributing Authors

Madelief Bertens m.g.b.c.bertens@hhs.nl is senior lecturer for the bachelor's in applied communication and associated as senior researcher with Research group Healthy Lifestyle in a Supporting Environment at The Hague University of Applied Science (THUAS). She teaches research methods in communication science, behavior change and health communication and supervises students' bachelor's theses. Madelief Bertens has a degree in neurological psychology (Msc) and in cultural anthropology, specialized in medical anthropology (Msc). In 2008 she completed her PhD dissertation at Maastricht University on promoting sexual health for women with a migrant background. She has been involved in health promotion and intervention planner as a lecturer, trainer and adviser since 2001.

Mark Bos mark.bos@gmail.com is head of the department Education and Research at the Faculty of Applied Social Sciences and Law. He has worked as a (science) communication advisor, teacher and researcher. As an advisor he has worked for VU University and the Netherlands Forensic Institute. As a teacher he has taught at VU University, Leiden University and The Hague University of Applied Sciences in communication science in general and science communication specifically. At VU University he taught science journalism, among other courses. As a researcher, he has done a PhD in science communication, focussing on how adolescents make sense of ecogenomics.

Barbara Campany Barbara.campany@ghd.com is an industry specialist in stakeholder engagement and risk communication with over 25 years of experience across Australia and NZ. With a strong focus on streamlining environmental approval processes through a risk communication framework, Barbara has garnered much of her experience through high profile, and often controversial major mining and infrastructure projects where communities, stakeholder groups and organizations are often divided. Regarded as a thought leader and specialist by her peers in community and stakeholder management, Barbara regularly provides training to build internal team capacity in risk communication and outrage management for large government agencies and corporations delivering major projects. Barbara has worked across numerous sectors including energy and resources, education, linear infrastructure, water infrastructure, urban renewal and precinct development.

Craig Cormick Craig.cormick@thinkoutsidethe.com.au is an Australian science communicator with over 30 years' experience working at the intersection of practice, research and policy - predominantly around contentious technologies such as biotechnologies and nanotechnology. He has been widely published on drivers of attitudes to new technologies, including in publications of *Nature* and *Cell*. He has worked for several Australian Government agencies including CSIRO, the Department of Industry, Innovation and Science and Questacon, before setting up his own company ThinkOutsideThe. He has been President of the Australian Science Communicators and is a member of the Advisory Board on Education and Outreach for the Nobel Prize-winning Organisation for the Prohibition of Chemical Weapons. He has published over 25 books – his latest being *The Science of Communicating Science: the Ultimate Guide* (2019).

Liesbeth de Bakker e.p.h.m.debakker@uu.nl is lecturer in science communication at the master programme Science Education and Communication of Utrecht University. Within science communication, she focuses mainly on informal science education. She teaches at the bachelor's and master's level and develops her own courses and projects. She coordinates the students' product development internships and supervises many of these projects. From 2013-2017 she was involved in the EU-funded project SYNENERGENE, which was focused on facilitating public dialogue about synthetic biology. Liesbeth studied biology at Leiden University (NL) and science communication (postgraduate level) at Imperial College (UK). She worked in science journalism at Radio Netherlands the Dutch International

Service for 9 years before becoming a science communication lecturer. She was editor of the Dutch predecessor of this book: "Wetenschapscommunicatie, een kennisbasis (2014)".

Anne M. Dijkstra a.m.dijkstra@utwente.nl is an assistant professor in Science Communication at the University of Twente in the Netherlands. Prior to her academic work, she worked as a science communication advisor. Anne studies the changing science-society relationship from a communication perspective. Her research focuses on understanding this relationship and, in particular, roles of the public, as well as, roles of researchers. Studies often relate to emerging technologies such as nanotechnology, biotechnology or human enhancement. Key words: public participation, science communication, risk communication and responsible research and innovation. She is involved in European funded projects NUCLEUS (No 664932) and GoNano (No 768622). She was a visiting researcher at Newcastle University and a visitor at the Institute of Advanced Study at Durham University. She teaches courses related to the science-society relationship about science communication, science journalism, and responsible innovation. As a volunteer, she co-organizes public meetings for the Science Café Deventer. She was editor of the Dutch predecessor of this book: "Wetenschapscommunicatie, een kennisbasis (2014)".

Maria E. Fernández Maria.E.Fernandez@uth.tmc.edu is Professor of Health Promotion and Behavioral Sciences at The University of Texas Health Science Centre at Houston School of Public Health, and Director of the Centre for Health Promotion and Prevention Research (CHPPR). She is an expert in health promotion planning and implementation research and practice. She has developed and evaluated health promotion and communication interventions particularly in the area of non-communicable diseases with a focus on underserved populations and the development of interventions to reduce health inequities. She is an author of Planning Health Promotion Programs: An Intervention Mapping Approach (2016) and the Handbook of Community-based Participatory Research (2017), and contributed chapters to the recently published Implementation Science across the Cancer Control Continuum (2018).

Yaela Golumbic yaelago123@gmail.com is a science communication researcher, emphasizing on citizen science as a way for enhancing public engagement with science. She leads the Israeli citizen science initiative "Sensing the Air" for engaging citizens in air quality research, and the "Radon home survey" for self-monitoring of radon levels in private spaces. Yaela is a member of the Taking Citizen Science to Schools (TCSS) research excellence center, and is the director of engagement in CitizenScience.Asia association.

Eric A. Jensen eric@methodsinnovation.org is a social scientist with a global reputation in science communication evaluation. Jensen's track record includes 13 external grants, 40 articles in peer-reviewed journals such as *Public Understanding of Science, Conservation Letters* and *Nature*, 5 books by publishers such as Cambridge University Press and SAGE and numerous influential reports for government and non-profit organisations such as the UK government's Department for Digital, Culture, Media and Sport and Department for Environment, Food, & Rural Affairs, Arts Council England, the National Coordinating Centre for Public Engagement, Association of Science & Technology Centers and the World Association of Zoos and Aquariums. Jensen is a Senior Research Fellow at the Institute for Methods Innovation (methodsinnovation.org), Senior Data Scientist at Inscico (Institute for Science & Innovation Communication) and Associate Professor of Sociology at the University of Warwick. His PhD in sociology is from the University of Cambridge. Keywords: Impact; evaluation; evidence-based science communication.

Marina Joubert marinajoubert@sun.ac.za is a senior science communication researcher and lecturer at the Centre for Research on Evaluation, Science and Technology (CREST) at Stellenbosch University, South Africa. Her research interests focus on scientists' role in public communication of science, online interfaces between science and society and the changing policy environment for public communication of science in Africa. She is also interested in the communication of contested topics in science, in particular the vaccine debate, via social media, as well as the societal factors that shape public science engagement in African countries. Marina is part of a research group attached to a South African Research Chair in Science Communication and coordinator of Africa's first online science communication training programme.

Edwin Koster e.koster@vu.nl is associate professor at the Department of Philosophy, Faculty of Humanities of Vrije Universiteit Amsterdam. He teaches philosophical courses for students in the Life Sciences, Educational Studies and Social Sciences. He studied Mathematics, History and Philosophy of Science, Philosophy of Religion and Spanish History, Art and Culture in Amsterdam (VU University), Madrid (UCM) and Princeton (PTS). He graduated in Mathematics (1992) and Religious Studies (1995) from the VU University Amsterdam. His PhD thesis – in the fields of Philosophy of Science and Philosophy of Religion – was about Narratives, Rationality and Religion (VU, 2005, cum laude). He is director of the educational program of philosophical reflection for all students at the VU University of Amsterdam and program director of philosophical studies. His research interests include philosophical questions regarding science, education, narratives, film, and religion.

Frank Kupper f.kupper@vu.nl is an interdisciplinary researcher, performer and facilitator. Currently, he is an assistant professor in science communication and public engagement at the Department of Science Communication of the Athena Institute, Faculty of Science, Vrije Universiteit Amsterdam. Blending an STS and Arts & Design perspective, Frank works on the conceptual and methodological innovation of public engagement processes to shape meaningful conversations at various science-society interfaces. Keywords are reflection, dialogue and transformative change. He has been part of various research projects at national and EU level, focusing on public engagement and responsible research and innovation. At the moment, he is the coordinator of the EU project RETHINK that develops innovative ways to do open and reflexive science communication across Europe. His teaching revolves around the same spirit of openness and reflexivity in science-society interactions, in courses such as science journalism, science in dialogue, and ethics.

Anne M. Land-Zandstra a.m.land@biology.leidenuniv.nl is assistant professor in science communication at Leiden University. She teaches about informal science education, supervises students during internships and thesis writing, and coordinates the science communication program. Her research focuses on citizen science as a form of informal learning, with a focus on participants' motivations; and on the role of real objects in science museums.

She is co-chair of the EU COST action Citizen Science to promote creativity, scientific literacy, and innovation throughout Europe. She received her PhD in 2012 in the USA for a study on the impact of an informal science program on high school students' knowledge of and interest in science. Before that she worked at the science museum Museon, The Hague, as science educator/curator.

Joanne Leerlooijer joanne.leerlooijer@wur.nl is lecturer in communication at Wageningen University and Research in the Netherlands. She teaches about designing communication interventions and supervises students who conduct their master's or bachelor's thesis or do an internship. She has also been involved in education innovation, in the design and facilitation of courses in an online distance learning master program. Prior to her work in Wageningen, Joanne worked for 10 years as a researcher with Rutgers (Dutch knowledge center on sexuality) in the field of sexual and reproductive health and rights in African and Asian countries. At that time, she was also involved as a volunteer in Adopt a Goat, a Dutch foundation working on the empowerment of more than 2200 single teenage mothers in rural Uganda. She completed her PhD in Maastricht combining the research in projects of both Adopt a Goat and Rutgers.

Nokwanda Makunga makunga@sun.ac.za is an associate professor at the Department of Botany and Zoology, Stellenbosch University and her research combines the areas of biotechnology, ethnopharmacology and phytochemistry. She is the group leader in medicinal plant biotechnology and her research is focused on medicinal plants, their cultural significance and opportunities presented for socioeconomic development. She teaches botany (plant physiology, chemical diversity in plants, ethnobotany and ethnopharmacology) at both the undergraduate and postgraduate levels. Her interests also lie in science communication targeted at improving the visibility of women in science and the public understanding of biotechnology and traditional medicines.

Henk Mulder h.a.j.mulder@rug.nl is senior lecturer in science communication and science & technology studies at the University of Groningen, The

Netherlands, and director of the Master Programme Science Education and Communication. He is also a co-ordinator of the Science Shop at the faculty of Science and Engineering. He teaches on risk communication and public engagement in various programs. Henk Mulder holds a Master in Chemistry and a PhD in Energy and Environmental Sciences. He has been involved in many European projects on Public Engagement, with a focus on research collaboration with civil society organizations; such as PERARES (which he coordinated) and Engage2020. He has published on nano-dialogues, policy-impact of engagement, and engagement in energy research.

Frank Nuijens frank@franknu.com was lecturer in science journalism at Delft University of Technology (NL) for 11 years, after a career as a science journalist and researcher at Dutch public television for 7 years, and as the editor-in-chief of the university newspaper and alumni magazine at Delft University of Technology for 7 years. He was also editor of the online science journalism course offered by the World Federation of Science Journalists. Currently, Frank is head of communications at ASTRON, the Netherlands Institute for Radio Astronomy, and he owns FrankNu Kenniscommunicatie, a company that offers training, workshops, advice and coaching to help scientists communicate their work. Frank holds a Master's degree in biology from Leiden University (NL).

Elaine Reynoso-Haynes elareyno@dgdc.unam.mx works in the DGDC where she is the director of Training and Research in Public Communication of Science. She has a degree in Physics and a Masters and PhD in Education. She coordinated the group of Planning (visitor studies) for the project "Science Museum *UNIVERSUM*" of which she was the first director from 1993 to 1998. She coordinates the Diplomado en Divulgación de la Ciencia (Post graduate 240 hour course in Science Communication) since 2007. Former president of the SOMEDICYT (Mexican Society for Science Communication from 2001-2003 and 2012-2014 and coordinator of the Northern Node of the RedPop (Latin American Network for the Popularization of Science and Technology) 2014-2016. She received the recognition for women academics "Sor Juana Inés de la Cruz" 2014 from the UNAM and the National Prize for Science Communication "Alejandra

Jaidar" in 2016. In 2019 she received RedPops Latin American Award for the Popularization of Science and Technology.

Frans van Dam f.w.vandam@uu.nl is lecturer in science communication and manager of education innovation at the Freudenthal Institute, Utrecht University. He initiated the development of the predecessor of this book, Wetenschapscommmunicatie, een kennisbasis (in Dutch), BoomLemma, Den Haag, 2014. In Utrecht, he co-coordinated an EU project aimed at integrating socio-scientific issues and inquiry-based learning in primary and secondary STEM education (www.parrise.eu, 2014-2017). Prior to his position in Utrecht, he worked as manager of an educational program on the biobased economy for Delft University of Technology. At Radboud University Nijmegen, Frans was head of communications for a national centre studying the societal aspects of the life sciences. At the Dutch consumers union, he has been policy advisor and lobbyist in the area of biotechnology, foods and farm animals. He co-organized the national debates on xenotransplantation (1999-2001) and genetically modified foods (2001).

Erwin van Rijswoud Erwin.van-rijswoud@oru.se is policy analyst at Örebro University in Sweden, Office for Academic policy. As quality coordinator he is primarily responsible for the internal reviews of educational programs. He holds a master degree in the history and philosophy of science from Utrecht University, and a PhD in social studies of science from Radboud University in the Netherlands. Before he was appointed to Örebro University, he worked in the Netherlands as a project manager for quality assurance of education and research, and as an assistant professor. His research and education dealt primarily with science communication, technology assessment and the role of experts in policy development and communication to the general public.

Roald P. Verhoeff r.p.verhoeff@uu.nl is assistant professor in Science Education at the Freudenthal Institute at the Science faculty of Utrecht University. He holds a Master's degree in Biology and a PhD in Biology Education and was involved in several projects which aimed to translate ethical quandaries on emerging science and technologies to science education. He has practiced and lectured on science communication and education for more than 12 years at Delft University of Technology, Radboud University and Utrecht University. His current research interest is in the integration of moral and societal implications of research in science education. As such, he is involved in the Erasmus+ INTEGRITY project in which he develops blended integrity education for PhD students.

Caroline Wehrmann c.wehrmann@tudelft.nl is assistant professor in Science Communication at Delft University of Technology in The Netherlands. She was one of the founders of the master *Communication Design for Innovation* at Delft University. 'Living labs' are part of the curriculum of this master program. In living labs students, researchers and private and public organizations collaborate to analyze and find solutions for complex socio-technological problems. Caroline investigates collaboration in living labs and in transdisciplinary teams to understand how these settings can contribute to gaining adaptive expertise: the ability to invent new approaches and solutions to problems that can't be solved on basis of routine. Key words in her research are: professional development, collaboration, co-creation, complex problems.

YIN Lin Yinlin213@126.com graduated from the Graduate School of the Chinese Academy of Social Sciences and received her LittD in 2005. In the same year, she began her career in science communication research at the China Research Institute for Science Popularization (CRISP), a national research institute devoting itself to both theoretical and applied research in science communication. Currently, she is an associate researcher and Deputy Director of the Division of Science Popularization Policy Research at CRISP. Yin Lin's research interests lie in the following fields: the history of science communication, popular science writing and publications; the mobilization of scientists in science engagement; RRI; and science culture in different social contexts. Either as team leader or core team member, she has been involved in more than 40 research projects in science communication. She has published, either in China or abroad, more than 30 peer-reviewed papers and book chapters, as well as 10 reports/proceedings.

Chapter 1

Setting the Scene

Anne M. Dijkstra, Liesbeth de Bakker, Frans van Dam,
and Eric A. Jensen

1.1 Introduction

Science communication is at the heart of many of the 21ˢᵗ century's most consequential issues. From climate change to artificial intelligence and biomedicine, science and technology are playing an important role in people's lives to an ever-greater extent. Science and technology are also considered important drivers for enhancing innovation. Moreover, citizens' role in engaging in democratic decisions about science and technology is vital, as such developments affect all people. This important role of science and technology leads to questions such as the following: How do people make sense of scientific and technological developments? How can societal needs and concerns be included when developing science and technology? How should communication about science and technology be conducted? Science communication practice and research is on the front line, helping both scientists and citizens grapple with such questions.

Communicating about science and technology comes in many different forms. Telling people about science is one important task. In addition, it is widely accepted that people should be able to engage with science and technology topics at a democratic level because science and technology affects all our lives. Communications on science and technology have been ongoing for a long time and have gained importance in recent years. Yet, science communication as a profession and a field of study is still relatively young. Historical events, societal changes, and other fields of practice and research have influenced the development of science communication. This book aims to provide readers with an accessible starting point to get an overview and to understand better what is known about science communication in practice and in research.

This book evolved out of a Dutch introductory text to science communication for Dutch practitioners and students. In recognition that science communication has become a worldwide practice and research field, this book has aimed to increase its international scope and relevance. An international review panel with well-respected colleagues from South Africa, China, and Mexico were asked for their guidance and input to extend the book's perspective. The book, hence, attempts to provide insights not only from a Western perspective. Instead, it includes a broader set of findings, principles, difficulties, and approaches that can flexibly be used to understand science communication in different cultural contexts and situations. This chapter sets the scene for engaging with science communication as a topic. It provides important concepts, ideas, and developments in science communication, which are presented within the context of a changing world, to aid in understanding the chapters that follow.

1.2 Science Communication: An Evolving Profession and Field of Study

Over the past few decades, especially since the 1980s, in many countries around the world, science communication has grown into an increasingly recognized profession and a field of study (see also Bucchi & Trench, 2016; Guenther & Joubert, 2017). Science communication always involves connections between science, technology, and society about (an application of) this science and technology. A great diversity of participants may be involved in this process, including scientists, policy-makers, activists, ordinary citizens, and other groups. The science communication process is dynamic, constantly changing, and driven by a variety of interpretations, views of science, and communication goals.

Science communication is a term that is widely used and interpreted in various ways. For this book, the editors have prepared a working definition of science communication, based, among others, on the discussion about public engagement from the website of the UK National Coordinating Centre for Public Engagement (2019):

> Science communication describes the many ways in which the process, outcomes, and implications of the sciences — broadly defined — can be shared or discussed with audiences. Science communication involves interaction, with the goal of interpreting scientific or technical developments or discussing issues with a scientific or technical dimension.

Approaches to science communication can range from an informative program on television in which information is transmitted to an audience (a so-called transmission-oriented activity) to dialogue sessions gathering public input about their views which will be strongly based on interaction between two involved parties (a so-called transaction-oriented activity). In transmission-oriented activities, one-way communication is mainly involved, while in transaction-oriented activities two-way communication is key. Goals for science communication may vary and overlap. They range from raising awareness and increasing appreciation for science and technology; sharing findings and excitement and, thus, aiming for enjoyment for science and technology; increasing non-scientists' knowledge and understanding; and influencing science-related opinions, views, and behavior or even people's policy preferences to engaging with others in order to include their perspectives in decisions about science and technology. The need for such a 'listening' approach in the last goal is particularly recognized with controversial science and technology topics (see also the report of the National Academy of Sciences, NAS, 2017).

In all science communication efforts and activities, science communicators take different roles, related to their aims (Jensen & Holliman, 2016). These roles can play out in transmission-based as well as in transaction-based approaches, or in anything in between. A journalist who writes a critical piece in the newspaper may aim to influence opinions, a museum staff member who develops activities for high school students may want to increase scientific understanding, while a scientist who presents an enthusiastic story for a science café public may be aiming for increasing knowledge and awareness. These are all examples of people who communicate about science and technology. And they all take up different roles in the communication process, as an intermediary, educator, facilitator, or expert. They are all practicing science communication. In addition to the roles these science communication professionals play, there is also a role for science communication researchers, that is, those who conduct research on science communication. These researchers or scholars often aim to better understand science communication processes as well as the effects of science communication.

As a field of study, science communication is heavily influenced by other disciplines, as shown in Box 1.1, which means that science communication practitioners as well as researchers bring in a rich variety of knowledge, related to their own backgrounds. The variety of communication approaches and roles for communicators, as well as their different backgrounds, make the field of science communication complex, challenging, and interesting.

Box 1.1:   Research disciplines and the field of science communication.

The field of science communication has been affected by several long-standing academic domains, most importantly by communication sciences, social studies of science and technology, (science) education sciences, and the natural sciences (Mulder, Longnecker & Davis, 2008). Important insights out of the academic domains of sociology and psychology feed three of these four key domains, and for the sake of overview, the domain of journalism and media studies is seen as a part of communication sciences.

Existing science communication courses taught at universities across the world often combine knowledge from several academic disciplines within their own curriculum along with the perspectives of practitioners (see Figure 1.1). According to Mulder, Longnecker & Davis (2008), knowledge from the natural sciences and the life sciences plays an important role in the 'translation' of information. Communication theories and communication skills provide a link between theory and practice. Knowledge about learning and teaching is also important in successful communication. This is especially the case in the area of informal learning. And the field of science and technology studies contributes to science communication through research into the interaction between science and society, its advice to policy-makers, and the reflective questions it raises about the nature and role of science and technology. Mulder, Longnecker & Davis (2008) also recognized other key knowledge domains such as sociology and psychology, and journalism and media studies.

Box 1.1: (*Continued*)

Science communicators can carry out their jobs informed by some of the different knowledge domains. For example, a museum staff member may use knowledge from the educational domain. A communication consultant at a hospital may use scientific knowledge about diseases and knowledge from communication studies when designing communication processes. Science domains that nourish science communication are themselves also influenced by developments in the science communication field. For example, natural scientists are becoming increasingly aware that scientific and technological developments are closely tied to social developments, and, therefore, in many countries, communication skills are now part of the expected competences for scientists (Gibbons, 1999). Over the past few decades, it has become increasingly accepted that multiple groups are involved in the complex relationship between science, technology, and society and the development of science and technology.

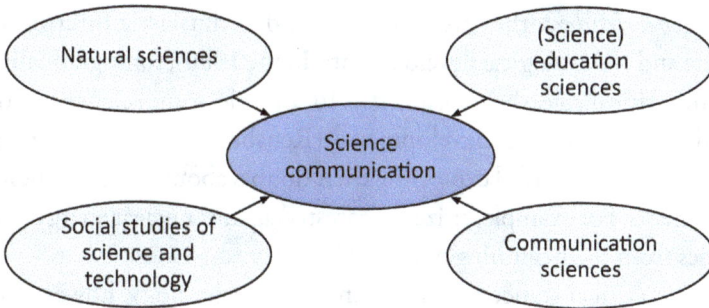

Source: Based on Mulder, Longnecker & Davis (2008).

Figure 1.1:   Research disciplines influencing science communication.

1.3 Changing Views on Science Communication

Starting in the early 19th century, in 1825, the physicist Michael Faraday, a member of the scientific society called the Royal Institution of Great Britain, initiated the annual Christmas lectures in which scientists presented scientific subjects to a general audience. These lectures continue to the present day and reach a large, mainly young audience. The Christmas lectures have often focused on the beneficial side of science and technology. However, critique of science and technology has grown, and nowadays, in many countries all over the world, subjects such as climate change, biotechnology, and vaccination are debated in the public domain. Informed and empowered citizens often criticize or at least doubt whether the outcomes of science and technology are set in stone.

A few events can be identified as game-changers along the pathway to increased democratic engagement with science. For one, Rachel Carson published her book *Silent Spring* in 1962, in which she criticizes the use of pesticides. In 1968, concerned scientists worried about the future of the world and founded the Club of Rome. They published the book *Limits to Growth* a few years later. Environmental awareness awakened in certain parts of the world. About a decade later, in 1979, nuclear energy was discredited by a leaking nuclear power plant at Three Mile Island in the United States.

These — and similar — events, first of all, opened the eyes of many science journalists. After the Second World War, science journalists often acted as *cheerleaders*; they were positive and enthusiastic interpreters of scientific and technological developments. In the 1960s, some gradually took up a more critical role, that of *watchdog*. In this role, some science journalists critically commented on developments (Rensberger, 2009). Informed by the media, publics started expressing their doubts about some technological developments. For example, citizens protested against nuclear energy in many countries from the beginning of the 1980s.

Such a critical stance toward science and technology, however, is not completely new. People have, for example, worried about the changes that trains would bring along in the beginning of the 19th century. And, as early as in 1663, the first cases were reported where workers destroyed textile machines out of fear of the technology and its implications for their lives.

The increasing resistance toward subjects such as GM foods made governments aware of the possible adverse economic consequences of rejecting new technologies. Therefore, government and policy-makers increasingly emphasized the importance of knowledge about science and technology. More knowledge and education, it was assumed, should make citizens adequately knowledgeable about science, *scientifically literate*, and would lead to more appreciation of science and technology and its products

(Bauer, Allum & Miller, 2007). It is this premise that defines the so-called *deficit model* of science communication. In this model, the communication process is defined as a one-way transmission (Nisbet & Scheufele, 2009), where greater knowledge leads to greater support for science, technology, and the institutional view of science.

The assumption that providing scientific information will lead to a positive appreciation of scientific and technological applications has turned out to be inaccurate for some topics and categories of people. In certain cases, studies have shown that people became more critical after receiving additional information. For example, this was found in public responses to genetically modified crops during a period of high-profile public debate on the topic in the UK (Marris *et al.*, 2001).

Informing people better, and thus increasing the public's scientific understanding is not viewed as a sufficient aim by many in the contemporary science communication field. Science communication is, or needs to be, more nuanced than simply telling the facts or telling the facts better (Bauer *et al.*, 2007). Moreover, science and technology will always be understood within their broader social context and, therefore, non-scientific factors play a role in science communication. Scientific information can often be interpreted in various ways (NAS, 2017), as is exemplified with knowledge about climate change. Furthermore, communicating about science is often mediated by others than scientists themselves, while people will judge information based on other factors such as their trust in the source, their existing knowledge, and their beliefs and values (NAS, 2017). Moreover, experts and citizens often perceive risks and benefits of science and technology differently.

In the early 1990s, social scientists argued for more openness and dialogue in the relationship between science, technology, and society with increased success in gaining interest in this perspective from policy-makers and scientific institutions in Europe. Dialogue and participation were considered a new approach aimed at restoring trust in science and technology (Bauer, Allum & Miller, 2007; Sturgis & Allum, 2004; Wilsdon & Willis, 2004). Public debates, organized in various European countries, tried to bring into practice such an open dialogue. For example, in the Netherlands, in the 1990s and the beginning of the 2000s at least five public debates about topics such as cloning (Dolly, the sheep), genetically modified food, and biotechnology were organized. The interactions between science and society, however, were not always implemented as intended by the social scientists who advocated this approach. Some of these dialogues were rather premeditated discussions where experts decided what to talk about and with whom. In turn, citizens did not always accept these public dialogue exercises and the desired high numbers of active participation in these debates were often not achieved (Dijkstra, 2008).

Since this initial burst of enthusiasm for public dialogue with science in Europe, initiatives that explicitly take public perspectives and values into account have continued to develop and gained ground in institutional and government policies in many countries. Accordingly, language within policy documents and funding schemes in many countries and at the European level moved from *public awareness of science* to *citizen engagement* and from *science and society* to *science in society* (Irwin *et al.*, 2018), or even *society with and for science* (European Commission, 2019). Aided by new technology, such as smartphones, citizens can now become data and knowledge producers as well, and scale-up the existing science communication initiatives such as citizen science, in which large groups of laypeople are involved in the process of doing research, or in helping set research agendas.

1.4 Science Communication in an Increasingly Changing and Global World

Science communication is always embedded in a wider social and cultural context. When changes occur, either at a local or global level, in the ways in which people communicate, learn and grow, and live together, then all these small changes are bound to impact science communication and shape it as a field of practice and scholarship. Hence, the call to understand science communication within the system it operates (NAS, 2017). This section provides an overview of important global developments for science communication which relate, first, to the content of science communication, thus science and technology information and knowledge; second, to the people involved in science communication; and, finally, to the communication means and approaches used.

1.4.1 *The Content*

Over the past few decades, huge changes have taken place in science and technology, including increased specialization and interdisciplinary working (Agar, 2012). In some fields, science has rapidly become a team effort. An extreme example concerns the publishing of research articles, with 1,000 or more contributions that are becoming more common in the field of particle physics (Mallapaty, 2018). In some countries, the focus in science and technology research gradually is shifting away from fundamental to more applied research. Economic exploitation of knowledge is promoted, and relevance of research for society is stressed. As a consequence, ethical and social aspects of new research fields become study objects as well. Partners other than researchers with their academic knowledge are asked to join projects in some contexts. Professionals and practitioners, for instance, nurses and farmers, can contribute with their professional knowledge. Members of the general public, sometimes called laypeople, can provide insights by sharing their local, experiential knowledge, for instance, their experiences as a patient or as an amateur geologist (Wynne, 1989). An increasing number of communication activities facilitate the participation of both scientists and layexperts as equal partners (Davies *et al.*, 2009).

Science and technology research is increasingly seen as a way to find solutions to the huge and complex problems that societies worldwide face. A large, complex problem with potentially far- reaching consequences is climate change. It is related to issues like feeding the world, resource depletion, and biodiversity. Innovative and sustainable solutions are called for, requiring the input of many different actors: scientists, professionals, and laypeople. This presents a huge challenge for science communication: how to motivate everybody to do the right thing; how to inform everybody effectively; how best to teach them the required skills; and how to engage with them? This may call for more elaborate and more specified communication approaches.

An aspect related very closely to science and technology is risk. Risk has become a more visible issue in recent decades. It plays an important role in heated debates, for instance, about genetically modified food, Universal Mobile Telecommunications System (UMTS) radiation, and climate change. According to sociologist Beck (1992), scientific risks play a crucial role in how contemporary society operates. He argued that the world is in a *risk society* phase, defined by the hazards that people live with each day such as nuclear weapons and climate change that were created by technological developments. More than before, people and institutions are aware of the risks facing them and demand that governments and industry take action. The credibility and expertise of scientists and technologists are essential for evaluating and understanding these risks.

1.4.2 *The People*

A very important societal change with widespread effect is that the nature of global economic activity has shifted toward greater technological development, thus increasing the global need for education and technical skills. Rates of education have increased globally. In Western countries, more people are gaining an academic education than ever before, while in developing countries more people are receiving basic education than before (UNESCO Report on Education, 2017).

The 21st century economy in the most advanced economies is increasingly based on digital and other non-physical goods and services and to a lesser extent on traditional physical products. This has necessitated a more educated workforce, and the proportion of university graduates has mushroomed in recent years accordingly (International Institute for Applied Systems Analysis, IIASA, 2014). In this context, formally recognized knowledge is key to economic success. Lifelong learning to help the population keep its knowledge up once they leave school is also important. In addition, technology and the instant availability of information online make it increasingly feasible for people to develop their own understanding of topics that were formerly the preserve of experts. This includes self-diagnosis and home-based medical diagnosis and patients taking increased responsibility for self-managing their health care.

Furthermore, in the Western world the role of the democratic citizen has increasingly been recognized in the context of science and technology policy. Citizens have become involved and engaged into dialogue about new developments, often science and technology related, that are about to take place. In different ways around the world, there have been initiatives to align priorities in science and technology with needs and values in society. For example, in the European Union, there has been an emphasis on developing a responsible approach to research and innovation through social inclusion, appropriate ethical consideration, public participation, open access, and other good practices within science.[1]

1.4.3 *The Means*

The Internet has greatly influenced science communication in the recent decades. It has enormously increased the amount of information available. From scientific programs on YouTube to the ever-increasing emphasis on open-access journal publishing, science information is at the fingertips of computer and smartphone users. The Internet has also 'democratized' information about science and technology by making it much more widely available and accessible to everybody. Yet, as a consequence of this exponential growth of available information, the focus in accessing knowledge has shifted from searching (just seeking out information) to sifting (separating good from bad information). In addition, Internet users need to learn how to work with new technology and deal with the overload of information.

Also, the onset of social media and its proliferation, Facebook, YouTube, Twitter, WhatsApp, Instagram, and other applications, urge people to prepare for a new communication era, that of online communication. This not only brings new opportunities for democratic engagement with science but also new challenges such as *filter bubbles*, where users are systematically fed information that aligns with their existing views (Jensen, 2011). New skills have to be learned, such as distinguishing real news from *fake news*. Such analytical skills are necessary to survive online. Some may even want to develop skills to become online information providers.

1.5 This Book

The complex changing relationship between science, technology, and society has caused science communication to develop as a field, from predominantly

---

[1] For example, www.rri-tools.eu/about-rri.

transmission-based activities to a mix of different approaches, which also include more transaction-based activities, such as a public dialogue about nanotechnology. In addition, the science communication field is affected by broad global developments such as science and technology becoming more specialized and tackling more complex problems, the drive toward higher education levels and democratization of societies, and the advent of the Internet and new communication tools such as social media. No standard approach exists for organizing science communication activities. Science communicators must design an approach that best fits the situation, the message, and the people involved. More in-depth insight into science communication processes and products will help both researchers as well as practitioners to undertake science communication activities more effectively.

As the science–society relationship is so complex, the chapters in this book address a variety of topics in an effort to enhance insights in science communication practice, research, and theory. The first four chapters introduce the field of science communication, while Chapters 5–9 provide insights into subdisciplines of science communication. These subdisciplines are by no means exhaustive but represent important fields of practice in science communication: informal science education, science journalism, risk communication, health communication, and environmental communication. The final chapter introduces research in science communication.

After this introductory chapter, which sets the scene, Chapter 2 sheds light on the core content of science communication: science itself. It presents different views of science which provide a basis for reflection on how science is constructed; its dependency on social, cultural, and economic contexts; and how such contexts influence the image of science portrayed by science communicators. The authors end the chapter with provocative questions that serve as a guide for this analysis.

Chapters 3 and 4 show how the field of science communication has become more complex in order to cover a wide range of motives to communicate about science with non-experts, with an increasing number of issues that must be addressed, the need for different models and strategies, new social responsibilities for science communicators, and new ways of relating with different sectors of society. The chapters address the actors or stakeholders in science communication. The discussions in these chapters offer a starting point for considering how to approach science communication.

Chapter 5 gives the reader an introduction to the field of informal science education, a field that is closely aligned with, and often overlaps with, science communication. People have the need to incorporate information, knowledge, and skills which are closely connected to science and technology. This has increased the demand for programs, activities, and settings for informal education as part of society's offer of lifelong learning opportunities.

Chapter 6 deals with the rapidly changing field of science journalism, its challenges, and its implications for science communication. This subdiscipline is most intensely influenced by the onset of Internet and the development of online communication and social media. In particular, these developments are relevant for the role of journalists and the framing of scientific information.

Chapters 7, 8, and 9 present three different contemporary subdisciplines of science communication: risk, health, and environmental communication, respectively. Even though the chapters deal with different content, they share a general approach in that much of the communication efforts in these domains are targeted toward attitudinal or behavioral change, be it focused on health, benefits and risks, or environmental sustainability. The communication strategies proposed in these three chapters — which are often related — are most useful in the analysis of how science communication can provide people with the necessary knowledge and tools to empower them. Authors in all three chapters address these topics, often emphasizing the individual level of communication.

Finally, Chapter 10 focuses on research and evaluation in science communication and includes a case on communicating about pseudoscience. As science communication becomes more and more professionalized, research evidence is becoming increasingly important to underpin the best practices in the field, while effective evaluation must be considered as a fundamental ingredient of the creative process of science communication initiatives. The chapter presents how research insights can help both researchers and practitioners.

This book is meant for professionals, students, and all those who look for an introduction into the quickly developing practice and discipline of science communication. By presenting a general overview of the science communication field with more in-depth insights into several subdomains, this book aims to provide an informative and enjoyable tour through the rich and varied field of science communication.

References

Agar, J. (2012). *Science in the Twentieth Century and Beyond.* Cambridge, UK: Polity Press.

Bauer, M. W., Allum, N., & Miller, S. (2007). What can we learn from 25 years of PUS survey research? Liberating and expanding the agenda. *Public Understanding of Science, 16*(1), 79–95.

Beck, U. (1992). *Risk Society. Towards a New Modernity.* London: Sage Publications.

Bucchi, M., & Trench, B. (2016). Science communication and science in society: A conceptual review in ten keywords. *Tecnoscienza (Italian Journal of Science & Technology Studies), 7*(2), 151–168.

Davies, S., McCallie, E., Simonsson, E., Lehr, J. L., & Duensing, S. (2009). Discussing dialogue: Perspectives on the value of science dialogue events that do not inform policy. *Public Understanding of Science, 18*(3), 338–353.

Dijkstra, A. M. (2008). *Of Publics and Science. How Publics Engage with Biotechnology and Genomics.* Enschede: University of Twente.

European Commission (2019). *Science with and for Society.* Retrieved May 20, 2019, from https://ec.europa.eu/programmes/horizon2020/en/h2020-section/science-and-society.

Gibbons, M. (1999). Science's new social contract with society. *Nature, 402,* C81–C84.

Guenther, L., & Joubert, M. (2017). Science communication as a field of research: Identifying trends, challenges and gaps by analysing research papers. *Journal of Science Communication, 16*(2), 1–19.

International Institute for Applied Systems Analysis (IIASA). (2014). *Education reconstruction for 1970–2000.* Retrieved March 31, 2019 from http://www.iiasa.ac.at/web/home/research/researchPrograms/WorldPopulation/Research/ForecastsProjections/DemographyGlobalHumanCapital/EducationReconstructionProjections/education_reconstruction_and_projections.html.

Irwin, A., Bucchi, M., Felt, U., Smallman, M., & Yearley, S. (2018). *Re-framing environmental communication: Engagement, understanding and action. Background paper* (The Swedish Foundation for Strategic Environmental Research). Retrieved March 31, 2019 from https://www.mistra.org/wp-content/uploads/2018/10/re-framing-environmentalcommunication_mistra-bp-2018-1.pdf.

Jensen, E. (2011). *Role of Social Media-based Public Dialogue.* UK Government report. Sciencewise. Retrieved March 31, 2019 from https://webarchive.nationalarchives.gov.uk/20170110135736/http://www.sciencewise-erc.org.uk/cms/assets/Uploads/Social-Media-Public-DialogueFINALPDF.pdf.

Jensen, E., & Holliman, R. (2016). Norms and values in UK science engagement practice. *International Journal of Science Education — Part B: Communication and Public Engagement, 6*(1): 68–88. doi: 10.1080/21548455.2014.995743.

Mallapaty, S. (2018). *Paper Authorship Goes Hyper. A Single Field is Behind the Rise of Thousand-author Papers.* NatureIndex. Retrieved December 20, 2018 from https://www.natureindex.com/news-blog/paper-authorship-goes-hyper.

Marris, C., Wynne, B., Simmons, P., & Weldon, S. (2001). *Public Perceptions of Agricultural Biotechnologies in Europe. Final Report of the PABE Research Project Funded by the Commission of European Communities* (nr. FAIR CT98-3844. DG12-SSMI).

Mulder, H. A. J., Longnecker, N., & Davis, L. S. (2008). The state of science communication programs at universities around the world. *Science Communication, 30*(2), 277–287. doi: 10.1177/1075547008324878.

National Academies of Sciences, Engineering, and Medicine (NAS) (2017). *Communicating Science Effectively. A Research Agenda.* National Academy of Sciences: Washington, DC. Retrieved March 31, 2019 from: https://www.nap.edu/catalog/23674/communicating-science-effectively-a-research-agenda.

National Coordinating Centre for Public Engagement (2019). Retrieved May 20, 2019, from https://www.publicengagement.ac.uk/about-engagement/what-public-engagement.

Nisbet, M. C., & Scheufele, D. A. (2009). What's next for science communication? Promising directions and lingering distractions. *American Journal of Botany, 96*(10), 1767–1778.

Rensberger, B. (2009). Science journalism: Too close for comfort. *Nature, 459*(7250), 1055–1056.

Sturgis, P., & Allum, N. (2004). Science in society: Re-evaluating the deficit model of public attitudes. *Public Understanding of Science, 13*(1), 55–74. doi: 10.1177/0963662504042690.

UNESCO Report on Education (2017). Retrieved March 31, 2019 from http://unesdoc.unesco.org/images/0024/002481/248136e.pdf.

Wilsdon, J., & Willis, R. (2004). *See-through Science. Why Public Engagement Needs to Move Upstream.* London: Demos.

Wynne, B. (1989). Sheepfarming after Chernobyl: A case study in communicating scientific information. *Environment, 31*(2), 10–15, 33–39.

Chapter 2

Views of Science

Edwin Koster and Frank Kupper

2.1 Introduction

Science is everywhere. From the smartphones in people's pockets to the yoghurt drinks with added minerals consumers can buy at the corner shop, to the medicine taken daily by grandmothers: all this would be unthinkable without scientific progress. The spectacular progress in science and technology since the beginning of the Enlightenment in the 17th century has scientized society in many ways. Science is still the driving force in society and an important part of culture. Agriculture, medicine, psychology, and technology are, for example, largely dependent on scientific knowledge. Government policy is also often based on scientific research. And that's not all. Scientific insights have led to, sometimes drastic, changes in human self-image. People appeal more and more to scientific knowledge to explain elements in everyday reality, such as stress, success, and being in love.

How people think about the relationship between science and the public and science communication ultimately depends on the underlying view of science they have. If science is viewed as neutral and objective, then people are inclined to communicate science in line with the transmission model (see also Chapter 1). If science is regarded as a social process that is influenced by various factors (such as culture, politics, and the economy), then people would communicate science in an alternative way. Various views of science exist, and professionals in science communication should be aware of this variety. The different views can be traced back to the work of philosophers in previous centuries. Some basic knowledge of philosophy of science is therefore indispensable for anyone who wants to communicate systematically and responsibly about science.

In this chapter, various views of science, described from a Western world perspective, will be discussed. It starts with a discussion of the so-called common-sense view of science, which is linked to the idea of science as a neutral and independent institution. This particular image of science dropped out of use in the last few decades of the 20th century, at least among philosophers. In connection with that, another image has come into play, whereby reference is made to the cultural influences that turn science into a social and value-laden process. As a consequence, the question whether science could still be considered as a reliable practice was raised.

Philosophy of science allows for at least two possible approaches to this important question. The first is that science is a self-purifying system that can guarantee public credibility due to specific feedback mechanisms and highlighted norms. The second is closely linked to the call for a socially robust science in which non-scientific perspectives and the concrete and local circumstances in which science and technology are developed and applied are represented. In this chapter, these approaches are introduced and critically scrutinized, and, last but not least, related to the field of science communication.

2.2 *Common Sense* as One View of Science

The particular view people have of science determines to a large extent their expectations of communication about science. Many people subscribe to the view that has become known as the common-sense view of science. This view consists of the idea that a scientist appeals to reasons and evidence or, to be more accurate, *logical deduction*[1] and empirically established facts when making an argument. Norms and values are of no consequence in this view of science. No external influences are thought to have any bearing on the results of scientific research. That is why science should be *autonomous* (politics and industry do not determine the direction of the research), *neutral* (scientists

---

[1] Logical deduction is a conclusion whereby a special statement is logically derived from a general statement. From 'all swans are white', it can be concluded that the swan that I will observe must be white.

are not guided by religious beliefs, political views, or financial interests), *independent* (moral judgments and ideological and other views and interests play no role in the acceptance of scientific research), *free of social value* (scientists are not involved in any way in how knowledge is applied), and *not normative* (scientists do not engage in an evaluation in terms of 'good' or 'bad' of the social consequences of the knowledge acquired by them). Due to the scientific method, the validity of hypotheses and theories can be determined and reliable knowledge can be formulated and applied to different areas in society (Koster, 2014a).

Prominent philosophers of science like Alfred J. Ayer (1910–1989) and Carl Gustav Hempel (1905–1997) endorsed the common-sense view (Ayer, 1962; Hempel, 2001; Koningsveld, 2006). Although they differ on a number of issues, both see science as a purely rational undertaking in which logical arguments and references to empirical proofs are primary. Making observations produces objective facts on the basis of which scientists can formulate laws and theories. This is called *induction*.[2] Next, they examine if these laws and theories can be confirmed by new observations. This process is called *confirmation*. According to this view, which is called positivism and can be regarded as a systematic elaboration of the common-sense view, a scientific argument is free of contradiction, and each step that is taken in such an argument follows logically from the previous step. These steps can be taken because objective facts justify it. In this process, scientists do not allow themselves to be influenced by their prejudices and beliefs. Science, in this view, is thus always about objective facts, and norms and values do not come into play.

Karl Popper (1902–1994) launched a fundamental critique of this so-called positivistic view of science (Popper, 1961 and 1963; Koningsveld, 2006). According to him, the empirical sciences should not be inductive in nature but *hypothetico-deductive*. In Popper's view, which is referred to as critical rationalism, it is impossible to make valid statements on the basis of limited information. Instead of the ideal of verification (e.g. confirmation), Popper formulated the *falsification* criterion. Hypothetico-deductive means

---

[2] Induction is an inference in which a general statement is derived from one or more particular statements. From a finite number of observations of white swans, I conclude that all swans are white.

that scientists will not emerge with a theory based on observations but, in a creative moment, first formulate a hypothesis, then deduce some predictions from that hypothesis, and finally test these predictions via observation. According to Popper, testing a hypothesis does not occur by searching for a confirmation, but by investigating whether the hypothesis can be rejected or not, the process of falsification. Science should be able to clash with the facts. Thus, if someone formulates the hypothesis, *All crows are black*, other scientists do not have to search for black crows (e.g. confirmation), but, for example, for a white crow (e.g. falsification). If no white crow is found after a thorough search, then one can, according to Popper, claim for the time being that the hypothesis is correct.

Despite the many differences between the positivism of Ayer and Hempel and the critical rationalism of Popper, their views are in agreement on important points. Both share, for example, the conviction that scientific theories based on rational argumentation and observed facts, in the absence of external influences, always provide better access to truth. The common-sense view of science presupposes this conviction and is in line with the ideas of both Ayer and Hempel on the one hand and those of Popper on the other hand.

The general structure of a scientific paper confirms the common-sense view. It reflects the idea of science as a purely rational enterprise. The authors present their own scientific research as a straightforward route, whereby a problem can be solved by following a meticulous method in a responsible way. The style and language of scientific articles strengthen this view; being characterized by a plain and straightforward style, clear and precise use of language, an accurate presentation of research results, the attempt to take logical steps, and to limit the story solely to the data that are relevant for the research. Literary language is avoided and technical terminology demanded. Scientists also eschew speaking in the first person. At most 'we' is used, but usually 'one' is the preferred term. The effect of this is that all emphasis lies on the content and not, for example, on the person of the scientist. Writing in the third person gives the impression that the researcher as a human being is of no relevance. It makes no difference who does the research. If it is done according to the right method, the results will be reliable.

The common-sense view is linked to the transmission model of science communication (see also Chapter 1). Science presents the truth about the world. That is why it seems logical for the communication professional, as a transmitter, to formulate a message for the passive receiver with the purpose of either informing or educating the latter. The relationship between science and public is asymmetrical here. In contrast to science, the public has no knowledge or power, and communication goes solely in one direction, from science to the public. The most important challenge for the communication professional, from the transmission perspective, would be to put the universal knowledge of science in the minds of an ignorant public in the most attractive and convincing form.

2.3 New Views of Science: Incidental Causes?

In the decades since the famous work of Popper, the common-sense view of science has been shaken in various ways. Cases of scientific fraud, as the recent case of Diederik Stapel in 2011, the Dutch social psychologist, who invented countless research facts, have contributed to this. The increasing collaboration with private enterprise has also led to doubts about the independence and disinterestedness of scientific research. One other factor that led to questioning these characteristics of the common-sense view of science was the discovery of publication bias in medicine. Various studies have shown that research with positive results regarding the effectiveness of medicines is much more often published than research with negative results. Consequently, the ideals of truth and reliability are strained. Are scientific results actually trustworthy and are scientists after the truth or only their own interests?

The possibility of individual scientists being wrong or making mistakes was recognized, before philosophers of science such as Ayer, Hempel, and Popper published their views. Already in 1942 the American sociologist Robert Merton (1910–2003) formulated four norms for preventing such incidents (Merton, 1942): *communism* (the results of scientific research must be publicly accessible), *universalism* (the assessment of scientific results must be independent of race, gender, social position, nationality, religious identity, and other factors deemed as irrelevant to the scientific process, *disinterestedness* (feelings of honor and personal interests must not have any influence on the results of research), and *organized skepticism* (a fundamental critical attitude toward scientific results is needed). In fact, Merton attempted to preserve the 'objectivity of *science*' by recognizing that the 'objectivity of *scientists*' is not guaranteed. Individual scientists should therefore be kept under close supervision to ensure the reliability of science as a whole. Merton's norms can be seen as an attempt to keep the ideals of the common-sense view of science intact and to see the cases of fraud and commercialization as nothing more than incidents that do not affect the objectivity of science as a whole (Radder, 2010b).

The question, however, is whether such cases are actually incidental or structural in nature. According to the common-sense view, science is not influenced by external factors like cultural background, social relationships, worldview, prestige, and financial interests. Historical, philosophical, and sociological studies in recent decades, however, have shown that precisely those factors have played an important role in many examples of leading scientific research.

A well-known example is the case of the so-called *Homo floresiensis* (Zimmer, 2005; Koster, 2014b; see Box 2.1). Two different groups of scientists defend opposite hypotheses on the basis of the same data. Their arguments are based on results of scientific research, but it is abundantly clear that they

---

Box 2.1: *Homo floresiensis* as a scientific scene of battle

The remains of *Homo floresiensis* were discovered in 2003: a small hominid that lived between 95,000 and 14,000 years ago. The fossils led to a major controversy. Peter Brown, one of the research leaders of the group that had the results published in *Nature*, claimed that it concerned a new hominid species, *Homo floresiensis*. It was claimed that this species descended from *Homo erectus*, and its small size was the result of the so-called *island effect*. Among other things, it possibly took over the use of stone tools from *Homo sapiens*. A hypothesis developed by the Indonesian anthropologist Teuku Jacob, however, claims that the skull was that of an ordinary *Homo sapiens* that had been deformed by disease, which is why it was so small.

The Achilles' heel for each of these views is the answer to the question: are the fossils representative of the average *Homo floresiensis* or do they constitute an exception, the remains of a deformed member of a species already known to have existed? From the perspective of the philosophy of science, it is interesting to see what arguments play a role in this discussion and how the issue is resolved. In other words, what are the data on the

---

Box 2.1: (*Continued*)

basis of which the paleontologists draw their conclusions, and what are the factors that make one of these views dominant?

Jacob claims that the so-called new human species was, in reality, a pygmy with a skull of extremely small dimensions as a result of a disease (*microcephalus*). Jacob has been doing archaeological research on Flores for years, digging into and analysing layers up to 10,000 years old without knowing that an archaeological goldmine lay a few layers deeper. Was this difficult to accept for him and did it lead him to oppose the research on *Homo floresiensis*? In 2006, Jacob and some other scientists published an article in *Proceedings of the National Academy of Sciences* (PNAS) in which they claimed that the so-called *Homo floresiensis* is not a new hominid species at all but a deformed *Homo sapiens*. Is Jacob's interpretation of the data informed by his own interests? Can he not live with the idea that he missed an easy target?

Peter Brown and the other researchers on the Australian team that discovered the remains of *Homo floresiensis* hoped, of course, that the findings they made would indeed lead to the conclusion that they had discovered a new hominid species. That would guarantee them eternal fame. That is why they speak about 'a few skeptics' who contradict their conclusions, thereby suggesting that a consensus about their hypothesis has been reached among the majority of scientists and dismissed views such as those of Jacob. They do give arguments for this dismissal, but these arguments are not compelling.

---

are influenced by the scientists' interests. These are interests that, in this case, involve prestige and status in the first place and money in the second. The acceptance of either hypothesis appears to depend on these factors as well (Figure 2.1).

A further question may be whether external factors, like reputation, only play a role in a case that involves a limited amount of data or, as some claim, that these external factors *always* influence the acceptance of scientific knowledge. In other words, is the interaction between science and external influences incidental or structural? And a follow-up question is: Is it possible and, if yes, necessary, to eliminate these external factors, whether they are incidental or structural?

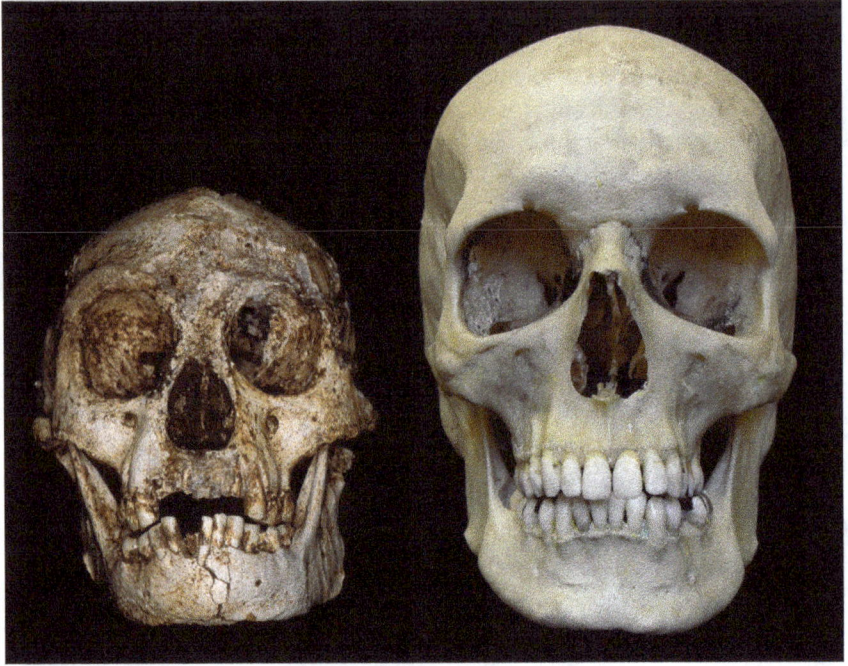

Figure 2.1: The skulls of *Homo floresiensis* and *Homo sapiens*. (Photo credit: Peter Brown.)

2.4 New Views of Science: Structural Causes?

Philosophers who have reflected on the phenomenon of science after Ayer, Hempel, and Popper, such as Thomas Kuhn (1922–1996), Harry Collins (1943–), and Bruno Latour (1947–), show that certain norms, values, and other external influences constitute an integral part of science. According to them, the common-sense view of science needs to be corrected. The following external factors, in addition to the internal factors of logic and testing by empirical facts, play a role in the acceptance of scientific knowledge, according to Kuhn, Collins, and Latour.

In *The Structure of Scientific Revolutions* (1970) Kuhn analyzes the history of physics. While the positivism of the common-sense view and the critical rationalism of Popper primarily *prescribe* how science should be done, Kuhn proceeds in a *descriptive* fashion. This step is generally designated as the *historical turn* in the philosophy of science. Kuhn claims that the development of science is not a linear process of gradually increasing knowledge but is characterized by revolutionary upheavals. In fact, Kuhn discusses two such upheavals: first, the one entailing the shift from the Aristotelian philosophy of nature of the Middle Ages to the mechanical natural sciences of Galileo and Newton. And second, the one entailing the shift from the mechanical natural sciences to the quantum physics of Einstein's relativity theory.

Central to Kuhn's analysis is the concept of *paradigm*. He defines the concept in a broad sense as the whole of a (specific) scientific approach: methods, techniques, skills, values, beliefs, and assumptions on which there is a consensus within a scientific group. Natural science that is, for instance, based on the theories of Galileo and Newton thus forms a paradigm. Kuhn describes the meaning of the concept 'paradigm' also in terms of the so-called *exemplars* (model solutions). These exemplars are standard examples of good scientific practice. They function as models to answer all kinds of problems that can arise within a paradigm.

In times of revolution, the problems that the paradigm can no longer resolve, pile up. Because of brilliant minds like Newton and Einstein, an alternative paradigm becomes available. But how can one decide if (a theory in) the new paradigm should be preferred above (a theory in) the old paradigm? According to Kuhn, historical research shows that the choice between two paradigms is not exclusively subject to processes of verification and falsification, of logic and empirical data. Subjective considerations also play a role in science. Kuhn attempts to show via a number of examples that individual scientists arrive at different choices because of, among other things, their own personal and professional background. According to Kuhn, social and psychological factors are ultimately also co-determinative for the acceptance and rejection of a scientific theory. These factors, called *external* in the common-sense theory, also belong to a paradigm.

Kuhn's conclusion about the acceptance of scientific knowledge turns thinking about science and external factors on its head. Scientific knowledge is, according to him, not incidentally but structurally influenced by such factors. Following Kuhn, other philosophers of science constructed a new view of science in which culture, social structures, worldview, prestige, and financial interests all play a role. The idea that science and values are separate worlds seems to belong definitively to the past. Box 2.2 gives insight into how science and values are related.

Representatives of the view of science in which science and values interact with each other are, among others, Harry Collins and Bruno Latour. In a number of case studies, Collins (1985) has shown that experiments cannot help in (severe) controversies between scientists. Because scientists look at research results through different lenses or perspectives in which different values play a decisive role, no independent criteria can be used for assessing the result of a controversial issue. Consensus on the meaning of a so-called *crucial experiment*, an experiment which will end discussion about a controversial issue once and for all, thus appears to be virtually impossible.

That is why the experts in question, in a debate on the validity of the results of an experiment, rely on judging the competence of the one carrying out the experiment. Unfortunately, no consensus can be reached on that issue either because usually the question who qualifies as an expert, and who does not, is not easy to answer. The discussion on *Homo floresiensis* (see Box 2.1) is one example of a controversy that reached an impasse because of a lack of

---

Box 2.2: Science and values

Regarding the subject of 'science and values', at least two important distinctions have been made. The first distinction is the one between epistemic and non-epistemic values (McMullin, 1983). It refers to *the kind of values* involved in science. Epistemic values are considered to be central to science and are regarded to play a legitimate role in the development of scientific knowledge. Thomas Kuhn mentioned five such epistemic values: accuracy, consistency, scope, simplicity, and fertility (Kuhn, 1977). Thus, according to Kuhn, if two competitive theories both cover the facts, the most accurate one deserves our preference, assuming that the theories have similar scores on the other four epistemic values. Which list of epistemic values plays a role in science is being debated — should 'explanatory power', for instance, be added to the five epistemic values mentioned by Kuhn (e.g. Lacey, 1999)? However, philosophers of science agree that epistemic values legitimately belong to science.

A more debated question is whether non-epistemic values are also involved in scientific research. Non-epistemic values include, for instance,

---

Box 2.2: *(Continued)*

personal values based on one's worldview or (financial) interests and also cultural, moral, economic, and political values. Regarding non-epistemic values, a variety of claims are formulated and defended. Some philosophers of science argue that the presence of these values in science should be avoided (e.g. McMullin, 1983), while opponents of this view try to show that non-epistemic values cannot be eliminated from scientific research, neither in practice nor in principle (e.g. Douglas, 2009).

To show *in which way* epistemic and non-epistemic values can play a legitimate or illegitimate role in science, Helen Longino introduces another distinction, namely between constitutive and contextual values (Longino, 1990: 4–7). Constitutive values are related to the aim of science and are necessary for the activity of science. Longino calls these values 'constitutive', to point out that they are the source of the rules determining what establishes acceptable scientific practice or acceptable scientific method. Contextual values belong to the social and cultural environment of scientific activities. They may influence the practice of science, but they are not necessary to conduct science. Constitutive values, for instance, determine in which way new medicines can be developed and tested, while contextual values can play a decisive role in answering the question of which medicine will be developed, possibly depending on the need for a certain medicine and the expected profit of its production.

According to Longino, contextual values do not only empirically shape scientific theories. Scientific practices and content even require the interaction with contextual values (Longino, 1990: 5). The latter, however, does not imply that science is hopelessly subjective. Due to the possibility of intersubjective criticism, it can be defended that science still is characterized by a kind of objectivity (Longino, 1990: 62–82).

---

unambiguous evidence. Collins holds that such controversies can only be resolved if social factors — such as statements by a scientist held to be an authority or the (professional) interests of the scientists — become part of the debate. What is called *external influence* in the common-sense view turns out, in this new view, to be *intrinsic* to the practice of scientific research.

Latour goes one step further (Latour & Woolgar, 1979; Swierstra, 2005). His work speaks of an *empirical turn* in the philosophy of science. As an anthropologist, Latour carries out participatory studies among scientists in a lab. The result is that, in his view, scientific knowledge occurs in a force field, in a messy struggle in which everything revolves around convincing others. For Latour, the others are no longer exclusively other scientists but also politicians, economists, religious leaders, and citizens. The force field shows the interconnectedness of science, technology, and society. The knowledge obtained rests on all kinds of factors in the force field, such as factors that are traditionally viewed as part of science (for example, fellow scientists, theories, equipment, statistics, measuring apparatuses, research assistants, and maintenance engineers) as well as non-scientific factors (such as culturally determined hierarchies, political constellations, and economic conditions). Within this force field, scientists try to establish supposedly true knowledge, knowledge that has eliminated all counterforces and is accepted by everyone. To win this power struggle, scientists need allies who can be acquired by convincing other scientists through negotiation and compromise and acquiring support in the world of politics and economics.

Ideas found in the work of Kuhn, Collins, and Latour are expressed in the transaction or dialogue model of the communication of knowledge (see also Chapter 4). If all knowledge is provisional and is the result of social and political processes, then subjective judgments will always be part of it. Scientists are not the only ones allowed to make judgments. Also, society should be able to express itself about the choices that are made in the scientific process and the values and assumptions lying at the foundation of the judgment. In the transaction model, scientists, politicians, journalists, other social players, and the public as a whole engage in dialogue with one another in order to arrive at new insights, alternative definitions of the problem, and solutions. It is crucial in this process that all sides be open to one another.

2.5 The Contextualization of Science

The analyses done by Kuhn, Collins, and Latour show how external factors influence the production of scientific knowledge. Following them, various sociologists and philosophers of science have attempted to map the interactions in scientific research since the 1990s. Many of these authors have established that scientific practice has changed radically over the course of the 20th century: science, politics, and society are no longer the isolated worlds they were once thought to be, but are increasingly interwoven with one another.

The question, of course, is whether a historical shift has actually occurred or whether the border between science and society has ever really existed. It is clear that the *view or image* of science as a sanctuary for the independent search for truth is still widely held, whereas the border between science and society in actuality is not as clear and undisputed. That is expressed, for example, in the formulation of research programs where, more than previously, the emphasis lies on the strategic objectives of the research, the production of relevant knowledge, a more intensive collaboration with industry, and an increase in the influence of social organizations on the agenda of the research. Since the year 2000, several researchers have described this transformation of the system of science by concepts like *mode-2 science, the triple helix, postnormal science,* and *postacademic science.*

The best-known and comprehensive concept is probably that of *mode-2 science* coined by the sociologists Nowotny, Scott, and Gibbons (2001). The central idea here is that the development of science occurs 'in the context' of practical applications. Especially in connection with new, upcoming technologies, such as information technology and nanotechnology, scientists often collaborate with other players, such as businesses and social organizations. This collaboration results in a dynamic interaction between various scientific and non-scientific disciplines. According to the mode-2 hypothesis, the classical image of scientific research as a whole of specialist disciplines working in isolation from each other and from society will slowly give way to another image in which a large diversity of people, such as scientists, professionals, and societal stakeholders, such as social organizations, and industry, are involved in the practice of science. This new image is called the *contextualization of science.*

The contextualization of science goes hand in hand with its economization. Economic thinking is increasingly penetrating the world of science: education is being evaluated more and more in terms of its return of interest, the business plan is invading the academy, scientific results are being viewed as consumer goods, and research institutions are acting like commercial enterprises. This leads to other choices regarding the direction and shape of research (Radder, 2010a). The agendas of scientific institutions, for instance, are very much influenced by the amount and impact of publications in international journals. Such results are used for hiring, promoting, and giving tenure positions to scientists. The increasingly stronger (and more international) competition makes scientists more inclined to investigate new research questions than to reproduce or replicate one another's experiments. In addition, registering the results achieved in patents limits the Mertonian striving to share knowledge freely. Patents do arrange for knowledge to be shared, but at the same time create a monopoly position. The increased time pressure and the pressure to publish make it more difficult and unappealing for scientists to check one another's work. The economization of science has also resulted in an increase in the collaboration between science and industry. While industrial and university laboratories developed their own practice and methods independently of each other for most of the 20th century, public–private collaboration is currently an increasingly frequent requirement for the financing of scientific research. These developments sometimes lead to a conflict of interest in science, as is exemplified in Box 2.3. Overall, with respect to the image of contemporary science, the different stakeholders have gained more understanding of, and sensitivity to, the fact that science develops in interaction with the social context.

Box 2.3: Conflict of interest in science

The blurred line between science and industry can easily lead to a conflict of interest. The publication bias in medicine is an example of this. Positive results of studies on the effects of medicines are more often published than negative results. Another example of the result of conflict of interest is *disease mongering*, a form of medicalization whereby the boundaries of a treatable disease are stretched to increase the market for medicines and aids. Alliances between pharmaceutical companies, doctors, and patient groups use the media to emphasize certain conditions and risk factors, with the result that biased views emerge with respect to the prevalence and seriousness of these conditions and factors.

One form of disease mongering is the presentation of conditions that are not life-threatening or impairing, as symptoms of a disease.

Box 2.3: (*Continued*)

A well-known example of this is the launch of the hair growth stimulant Propecia by the pharmaceutical company Merck in Australia. The advertising campaign, which gained a large amount of media attention, suggested that baldness among men could lead to severe emotional trauma. According to a new 'study' by the *International Hair Study Institute*, one-third of men suffer from a form of hair loss that could, in the long run, have a negative impact on job security and well-being. It was later discovered that Merck was behind the foundation of this institute. This example shows how disease mongering can be used deliberately to realize economic interests.

Disease mongering can, however, also occur unknowingly. Companies, doctors, and patient groups can act with the best of intentions to improve human health, but, at the same time — and unintentionally — adversely affect that health through medicalization. An example of this is osteoporosis. Although slowing bone loss seems to decrease the risk of future fractures, many people have a very low risk of breaking something. Therefore, it is difficult to determine the necessity of preventive treatments for osteoporosis. In addition, many non-medical treatment alternatives are available, such as vitamins and exercise. Nevertheless, health organizations encourage menopausal women to have themselves tested by a doctor for osteoporosis or to use a home test. Preventive treatment allows existing risk factors to become conceptualized as a disease, and, thus, these factors are medicalized. Precisely, the fact that the mechanism of conflict of interest can also occur unknowingly, makes it all the more complex for the scientific system to protect itself against the unwanted influences of stakeholders.

2.6 Credibility and Trust

Is this contextualization of science paying a price in terms of credibility and public trust in science? In the 1990s, the issue of a trust crisis in science was the subject of much discussion. New, upcoming technologies, such as nuclear energy and biotechnology gave rise to social controversies and an outright confrontation between science and the public. In countries like the Netherlands and the United Kingdom, this created a discussion regarding the legitimacy of science (Irwin & Wynne, 1996; Levidow & Marris, 2001).

Current research into the issue of public trust in science portrays mixed messages. Surveys in the USA and EU, as well as individual EU countries such as Germany and the Netherlands, roughly show the same picture when it comes to trust in science (Funk, 2017; van den Broek-Honingh & De Jonge, 2018). In general, public confidence in science is relatively strong, at least compared to other societal institutions such as the government, industry, and media. Science as an institution is associated with important values such as neutrality and independence. In specific cases of controversial developments however, such as vaccination, climate change, and biotechnology, public trust is more varied. These developments lead to concerns about risks for people, animals, and the environment and give rise to fundamental questions about wider moral issues. Especially the interwovenness of scientific pursuit with the interests of government and industry may lead to a decrease in trust in those cases. Furthermore, citizens appear to be more skeptical toward individual scientists themselves, with respect to, for example, their honesty and reliability and the way scientific research is organized.

These developments, together with the altered relationship between science and society, mean that the image of scientists as reliable, neutral, and independently acting individuals is not self-evident anymore. The complexity of public trust in science makes the public credibility of contemporary scientists a matter of dispute. Terms such as *post-truth* and *fake news* are increasingly prevalent in the media. It urges scholars who study the phenomenon of science to rethink the science–society relationship. Generally speaking, the credibility of science can be regained in the eyes of the public in two ways. Some philosophers of science still view contemporary science as a self-purifying system that can adequately protect itself against the unwanted influence of external forces. Other philosophers see this as illusory. They argue that the new interwovenness of society and science requires an alternative system of quality control and new forms of science that are 'socially robust'.

2.7 Science as a Self-Purifying System

One example of science as a self-purifying system is the result of the debate on *Homo floresiensis*. The struggle for the correct interpretation of the spectacular find on the Indonesian island of Flores involved two parties, generally speaking. One party asserted that the find pointed to a new hominid species, whereas the other claimed that the remains were those of a deformed *Homo sapiens*. The difference in interpretation was thought to be mainly a matter of dispute on competence and factors like status and prestige. Do these social factors really have the last word in the discussion on *Homo floresiensis*? That did not appear to be the case. Box 2.4 shows that over the course of time, various feedback mechanisms helped yield a definite answer.

In the *Homo floresiensis* case, more scientific support was found in favor of the discovery of a new hominid species through further thorough scientific analysis. The method of converging proof was used which together

---

Box 2.4: Looking at *Homo floresiensis* again

In 2007, *Science* published a study in which one of the conclusions was that the wrist bones of the skeleton deviated strongly from those of *Homo sapiens* and Neanderthals. The wrist bones of the former showed strong similarities with those of humanoid apes and early humanoid species. By analyzing the wrist bones as evidence along with the only skull discovered — the interpretation of which led to so much commotion among the researchers — the hypothesis of a deformed *Homo sapiens* was weakened. Supplementary 'converging proof' appeared in *The Journal of Human Evolution* (2009). A CT scan of the skull led to the conclusion that the brain of the *Homo floresiensis* did not resemble that of a *Homo sapiens* that had been deformed by *microcephalus*. A special issue of the same journal, later in 2009, published research results that show that *Homo floresiensis*, on the one hand, belonged to the hominid family. This was based on discovered characteristics like the thickness of the skull, the height of the face, and the form of the fibula. On the other hand, the bones had all kinds of primitive characteristics that were not even found in *Homo erectus*, who lived from about 2,000,000 to 400,000 years ago. Such characteristics included the shape of the jaw, the short calves, and the deviating wrist bones.

The hypothesis that the discovery of the remains could lead to the recognition of a new hominid — *Homo floresiensis* — is, thus, finding more support by the thorough analysis of new evidence. Because scientists looking for solutions via completely different approaches to and methods for the same problem, and along different paths, are reaching conclusions that reinforce one another, the reliability of scientific results is increasing.

with various feedback mechanisms forms the so-called *knowledge filter* (see Figure 2.2 for an illustration of the knowledge filter). The core idea of this filter, constructed by Henri Bauer (1992), is that the development of scientific knowledge can be viewed as an evolutionary process. Just as in nature, in science, as well, it is a matter of reproduction (there are carefully made copies of scientific results), variation (there are different approaches

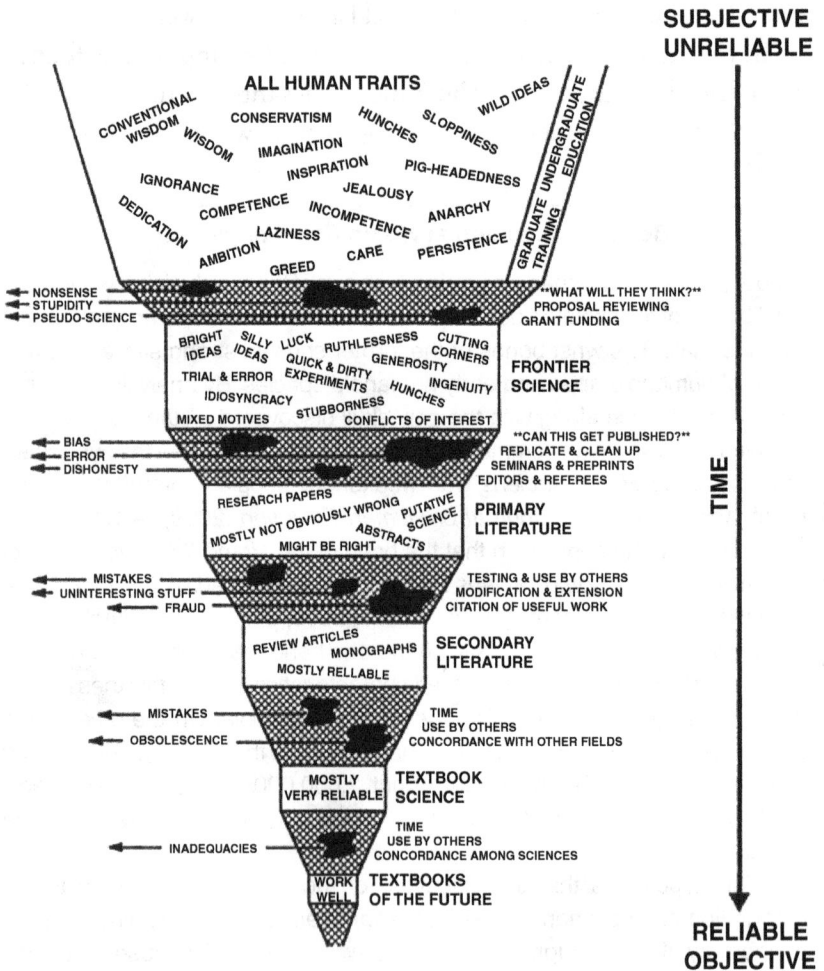

Source: Based on Bauer (1992).

Figure 2.2: The knowledge filter.

and solutions for similar problems), and selection (over time the best theories remain while less good theories are eliminated). The term 'best theories' in the knowledge filter concern theories on which the scientific community agrees. Only knowledge that is agreed upon within the forum of experts (ultimately) can claim (provisional) acceptance.

According to Bauer (1992), the agreement reached is to be based on the method of converging evidence, on various so-called *feedback mechanisms*. These mechanisms distinguish science from other areas like art, religion, and politics. Due to double-blind experiments, peer review, repetition of experiments (reproductions and replications), statistical tests, and agreement with other disciplines and sciences, rational consensus in science is possible. By applying these feedback mechanisms, results can be critically tested, corrected, supplemented, and reinforced.

To reinforce the self-purifying ability of science and to exercise damage control in relation to scientific integrity, research authorities across the world have developed principles of research integrity. These principles are made explicit in many different codes of conduct. Often, these codes refer to principles similar to Merton's ethos and are issued by national or international research associations, such as, for example, ALLEA (the association of All European Academies of Sciences) (ALLEA, 2017). In their code, ALLEA formulates four core principles of research integrity (honesty, accountability, fairness, and good stewardship). Guidelines for good research practice are derived from these principles and are aimed at preventing research misconduct. Often, research integrity is valued not only because it increases the scientific quality but also because it is argued to increase public trust.

2.8 Socially Robust Science

Is the idea that science can save or purify itself nonetheless an illusion, no matter how deeply it is cherished? One condition for the effect of the feedback mechanisms is, for example, that the researchers in question are independent of one another. If they are not, one can think here of the many examples of the development and testing of medicines by researchers from the same pharmaceutical company, then the feedback mechanisms threaten to degenerate into show procedures that mask personal and collective interests.

In addition, will scientists be able to meet the sharpened feedback mechanisms and the code of conduct? Just think of the discussions about sloppy science. Scientists are also human and are sometimes seduced by their own interests, needs, worries, or those of others.

Moreover, the knowledge filter implies that the system needs time for self-purification. Reliable results are part of *consensual science* or *textbook science* and not of *disputed science* or *frontier science* (Longino, 1990; Bauer, 1992). In the current media society, the circulation of information goes so quickly, however, that there is no time for a period of critical reflection before making the results public. Scientific news is broadcast on TV, radio, and the Internet every week, every day, every hour. Therefore, the media often make do with the very provisional results of frontier science that have only a limited 'half-life' (see also Chapter 6).

The researchers Functowicz and Ravetz (1993) pointed to yet another reason why the self-purifying mechanism of science takes too long. They argue that, in tackling complex problems like climate change, the classical means of doing science is no longer adequate. The complexity here is, namely, so large that researchers are very unsure about which facts are relevant for solving the problem. At the same time, so much is at stake that scientists simply cannot wait until a consensus has been reached via time-consuming feedback mechanisms. Functowicz and Ravetz therefore argue in these kinds of complex problem situations for what they call *postnormal science*. This is science in contrast to Kuhn's idea of normal science in episodes in which the prevailing consensus is not disputed. Postnormal science questions not only facts and methods but also starting points and values and broadens the decision-making on scientific questions to include the participation of non-scientific actors.

In practice, this would mean that the usual work of scientific advice councils would be expanded by a democratic process of discussion among a broad variety of stakeholders. The recognition of experiential knowledge, local knowledge, or lay knowledge as a surplus value for scientific processes can contribute to the social anchoring of these processes. Consultation between scientific and social 'experts' is expected to lead to a more balanced contribution of science to the development of society. Classical feedback mechanisms, such as the peer review system, should be supplemented by other methods of assessment in which the perspectives of all stakeholders are involved. The forum of experts should therefore be broadened into an *extended peer community* (Functowicz & Ravetz, 1993). The *extended peer review* has three aspects. First, it concerns the necessity to test new knowledge outside the laboratory. Second, this test must be done via the active involvement of a broad group of experts, including end users and laypeople. Finally, the contribution of these actors must be viewed as a process of permanent involvement (Koster, 2014c). In this way, science becomes more socially robust.

Socially robust science aims to open up research processes to the wishes and needs of various, and vulnerable, groups in society. Moreover, this democratization of science may profit from the practical knowledge of non-scientific actors in the approach to social problems (see also Chapter 4). Nonetheless, critical questions should be posed. Does the contextualization of science not lead inevitably to a loss of independence and objectivity? Does the distinction between science and politics disappear from this view of science? Certainly, due to unequal expertise, the risk emerges that support for certain decisions is given on the basis of political influence and rhetorical gifts, rather than through the careful weighing of information. And who then will finally reap the profits from the contextualization of science? Does it not unintentionally give free rein to the negative effects of economization? Will the further integration of science and society lead to more democracy or to its opposite: the hegemony of socio-economic factors, of which public–private partnerships, consultancies for social issues, and the 'enterprising university' are simply the precursors?

Indeed, although initiatives toward socially robust science have certainly opened up the scientific process to public scrutiny, the focus often remained on staging the discussion of ethical and societal aspects without actually changing the research practices themselves. The framework of Responsible Research and Innovation (RRI) aims to change this. It was introduced as a new framework for the governance of science aiming to integrate ethical reflection, public engagement, and responsive change (Stilgoe *et al.*, 2013). RRI seeks the development of platforms and processes that enable social actors to participate meaningfully in the research process from start to finish. The European Commission has embraced this inclusive and reflective approach to scientific research in its research funding policy for 2014–2020. RRI of course builds on a myriad of approaches to support the interaction of and facilitate the integration of science and society that have been developed in the past decades. A common thread in these approaches is the idea of a transformation of the world of science into a more open system that actively seeks and maintains connections with society and its societal needs, values, and concerns.

2.9 Conclusion

The views of science are changing. The demands on trustworthy science are changing as well. The production and acceptance of scientific knowledge is a result of internal factors, such as empirical evidence, logical reasoning, and argumentation, and external factors. External factors include cultural and political aspects, and economy and finances. So science does not happen in isolation but develops closely intertwined with technology, economy, politics, and society. It is set up as a self-purifying system so that the most trustworthy and valid knowledge can be produced. However, socialization and commercialization have limited the extent of its self-purifying capacity. So only if both internal and external factors are taken into account and dealt with in talking about the production and acceptance of scientific knowledge can trust be built.

Therefore, since the 1990s, an increasing call for public influence and participation in making decisions about the questions and directions of science has emerged. This resulted in a proliferation of participation experiments in various research areas such as biomedical research, sustainability research, and newly emerging technologies. A common goal of these experiments is to make the production of scientific knowledge socially robust. This means involving more non-scientific stakeholders and bringing other sources of knowledge and norms and values into the dialogue about science. Concerns about the impacts of knowledge should become an integral part of the production of the knowledge. Ideally, participation of diverse stakeholders should be facilitated early on or upstream in research processes, when choices about their direction can still be made.

A greater involvement of society in scientific research also requires attention for the communication processes that play a role here. Scientific and social actors come from different practices and cultures. This means that they have different perspectives on reality and use a different language to express what they think is important. Furthermore, these actors may have different values, interests, and views of science. The view of science determines in part the way in which science is communicated. The common-sense view of science often leads to a transmission approach. The socially robust science view leads to a transaction approach, such as dialogue formats. Productive dialogues between science and society, therefore, need to be professionally designed and facilitated. This is an important task for future communicators of science.

Acknowledgments

This publication was made possible through the support of a grant from Templeton World Charity Foundation. The opinions expressed in this publication are those of the authors and do not necessarily reflect the views of Templeton World Charity Foundation.

References

ALLEA (2017). *The European Code of Conduct for Research Integrity*. Berlin, Germany.

Ayer, A. J. (1962). *Language, Truth and Logic*. London: Gollanc.

Bauer, H. H. (1992). *Scientific Literacy and the Myth of the Scientific Method*. Urbana [etc.]: University of Illinois Press.

van den Broek-Honingh, N., & de Jonge, J. (2018). *Trust in Science in the Netherlands — Survey Monitor 2018*. The Hague: Rathenau Instituut.

Collins, H. (1985). *Changing Order. Replication and Induction in Scientific Practice*. London: Sage.

Douglas, H. (2009). *Science, Policy, and the Value-Free Ideal*. Pittsburgh, PA: University of Pittsburgh Press.

Functowicz, S. O., & Ravetz, J. R. (1993). Science for the post-Normal age. *Futures, 25*(7), 739–755.

Funk, C. (2017). Mixed messages about public trust in science. *Issues in Science and Technology, 34*(1), 86–88.

Hempel, C. G. (2001). *The Philosophy of Carl G. Hempel. Studies in Science, Explanation, and Rationality* (edited by J. H. Fetzer). Oxford: University Press.

Irwin, A., & Wynne, B. (eds.), (1996). *Misunderstanding Science? The Public Reconstruction of Science and Technology*, Cambridge: Cambridge University Press.

Koningsveld, H. (2006). *Het verschijnsel wetenschap. Een inleiding tot de wetenschapsfilosofie*. Amsterdam: Boom.

Koster, E. (2014a). Wat is wetenschap? Een introductie. In E. Koster (Red.), *Wat is wetenschap? Een filosofische inleiding voor levenswetenschappers en medici*. Amsterdam: VU University Press, pp. 9–31.

Koster, E. (2014b). De evolutie van de mens. Over wetenschap en waarden. In E. Koster (Red.), *Wat is wetenschap? Een filosofische inleiding voor levenswetenschappers en medici*. Amsterdam: VU University Press, pp. 47–67.

Koster, E. (2014c). *Science in Transition?* Kritische reflecties op hedendaagse wetenschap. In E. Koster (Red.), *Wat is wetenschap? Een filosofische inleiding voor levenswetenschappers en medici*. Amsterdam: VU University Press, pp. 303–328.

Kuhn, T. S. (1970). *The Structure of Scientific Revolutions* (2nd edn., Enlarged). Chicago: The University of Chicago Press.

Kuhn, T. S. (1977). Objectivity, value judgment, and theory choice. In T. S. Kuhn (ed.), *The Essential Tension. Selected Studies in Scientific Tradition and Change*. Chicago, IL: University of Chicago Press, pp. 320–339.

Lacey, H. (1999). *Is Science Value Free? Values and Scientific Understanding*. London: Routledge.

Latour, B., & Woolgar, S. (1979). *Laboratory Life: The Social Construction of Scientific Facts*. Beverly Hills, CA: Sage.

Levidow, L., & Marris, C. (2001). Science and governance in Europe: Lessons from the case of agricultural biotechnology. *Science and Public Policy*, 28(5), 345–360.

Longino, H. E. (1990). *Science as Social Knowledge. Values and Objectivity in Scientific Inquiry*. Princeton, NJ: Princeton University Press.

McMullin, E. (1983). Values in Science. In P. D. Asquith, & T. Nickles (eds.), *PSA 1982*, Vol. 2, East Lansing, MI: Philosophy of Science Association, pp. 3–28.

Merton, R. K. (1942). The normative structure of science. In N. W. Storer (ed.), *The Sociology of Science*. Chicago, MI: University of Chicago Press, pp. 267–278.

Nowotny, H., Scott, P., & Gibbons, M. (2001). *Re-thinking Science: Knowledge and the Public in an Age of Uncertainty*. Cambridge: Polity.

Popper, K. R. (1961). *The Logic of Scientific Discovery*. New York, NY: Basic Books.

Popper, K. R. (1963). *Conjectures and Refutations. The Growth of Scientific Knowledge*. London: Routledge and Kegan Paul.

Radder, H. (ed.), (2010a). *The Commodification of Academic Research. Science and the Modern University*. Pittsburgh, PA: University of Pittsburgh Press.

Radder, H. (2010b). Mertonian values, scientific norms, and the commodification of academic research. In H. Radder (ed.), *The Commodification of Academic Research. Science and the Modern University.* Pittsburgh, PA: University of Pittsburgh Press, pp. 231–258.

Stilgoe, J., Owen, R.J., Macnaghten, Ph. (2013). Developing a framework for responsible innovation. *Research Policy* 42(9), 1568–1580.

Swierstra, T. (2005). Bruno Latour. In M. Doorman, & H. Pott (eds.), *Filosofen van deze tijd.* Amsterdam: Bert Bakker, pp. 427–443.

Zimmer, C. (2005). *Smithsonian Intimate Guide to Human Origins.* New York, NY: HarperCollins.

Chapter 3

The Process of Communicating Science

Caroline Wehrmann and Anne M. Dijkstra

3.1 Introduction

Several developments have been decisive for the contemporary practice of science communication. Science communication is no longer a so-called translation of scientific information for the general public. It takes place in the dynamic environment of science and technology in which various actors with different interests try to reach a variety of goals by means of communication.

This chapter takes the perspective of a science communicator who is involved in the practice of communicating science. To work as a professional, it is good to understand the process of communication. Communication models can help to get some grip on communication processes. Over the years, various communication models — some basic, some more advanced — have been developed. As a general rule, all models may have some relevance because each of the models will highlight different aspects of the communication process.

Section 3.2 starts with discussing two communication models in more detail. Then, a set of important characteristics of science communication processes will be described. Based on the insights of those sections, the main building blocks in developing a science communication strategy will be explained. In the final paragraph, the question is raised as to what extent communication theories can be of help to build those strategies. A number of examples are provided to illustrate the added value of communication theories.

WHO	SAYS WHAT	IN WHICH CHANNEL	TO WHOM	WITH WHAT EFFECT?
COMMUNICATOR →	MESSAGE →	MEDIUM →	RECEIVER →	EFFECT

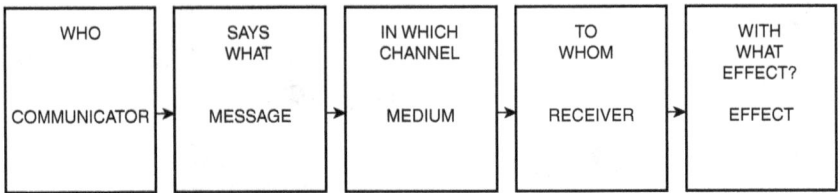

Source: Lasswell (1948).

Figure 3.1:   Lasswell's model of communication.

3.2 Models of the Communication Process

Communication models can help to understand the essence of the process of communicating science. Many basic models of communication are focusing on unilateral communication. An example is one of the oldest and probably the most well-known model of Lasswell, which dates back to 1948 (see Figure 3.1).

The model of Lasswell focuses on communication as a one-way communication process: it is a linear model. Still, it is relevant, because an important part of people's communication efforts are aimed at transmission. According to Lasswell's model, a communicator has to make a number of coherent choices in communicating to others. The model emphasizes that when a person (the sender) wants, for instance, to inform or persuade another person (the receiver), the sender must take a number of choices and align them: the sender has to take a clear stand on the message he or she wants to convey, must know who the desired recipient is, and has to tune the message and channel to this recipient. To obtain the desired effect, it is necessary that a communicator is found sufficiently credible by the receiver, that the message suits the receiver, and that the receiver is also receptive to it.

Communication is not only unilateral but can also be a process of interaction. Several communication models support the interaction model. Figure 3.2 is an example of such a model (Oomkes, 2013). For the sake of convenience, the model is based on one sender and one receiver. Obviously, in real life the possibilities are endless. After all, communication is happening in all sorts of settings, with various people.

The figure shows that *a source* (or sender) communicates departing from one's own knowledge and experience, standards, and preconceptions.

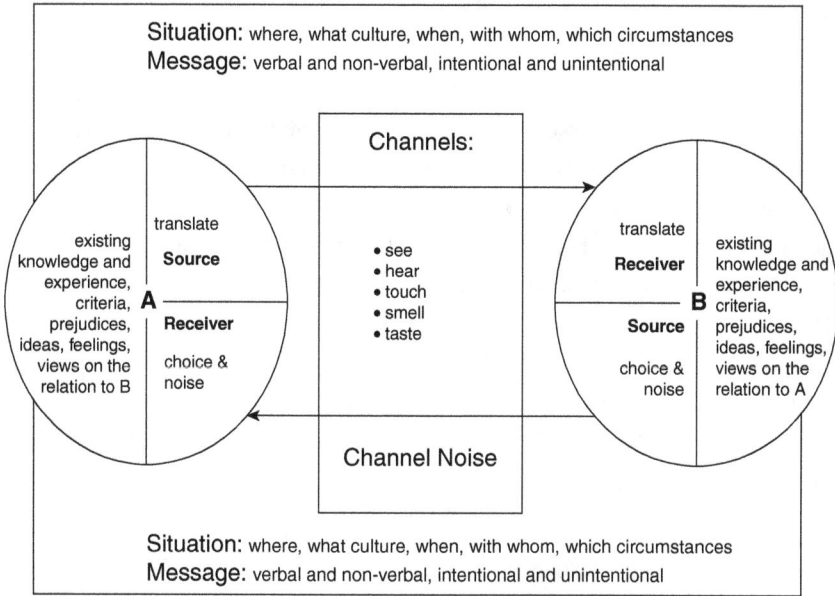

Source: Based on Oomkes (2013).

Figure 3.2: A model of communication from an interaction perspective.

But these are not the only factors that determine how a person chooses the message and the means of communication: the assessment of the situation and relationship with the *receiver* also play significant roles, whether consciously or subconsciously.

What often makes communication complicated is that the receiver interprets the *message* based on his or her own knowledge, experience, standards, and preconceptions as well. The *situation* in which communication takes place can also have a substantial effect on the way in which the message comes across. And there may be interference, which is described by 'anything that disrupts the communication'. Oomkes (2013) referred to two types of interference *internal noise*, which is when communication is disrupted by the internal state of one or more of the communication partners. When a person is not feeling well, for example, or when people have their mind on something else, they may miss parts of the message as a result. *External noise* concerns the message being disrupted by some other noise at the moment of the communication, which also can cause the receiver to miss certain parts of the message (see Figure 3.2).

From the perspective of the interaction model of communication, it is important to be open to each other. For effective interaction, participants must have an equal opportunity to take the initiative and to determine the content of the conversation. They must also be willing and able to listen to each other and to take the suggestions put forward by the other person into consideration. It is also important that discussion partners take each other seriously and respect each other. Another important criterion is that the topic of conversation should be allowed to change and develop as the discussion unfolds: people should feel free to give their point of view and know that they are being listened to. If two people have different but fixed opinions, there is no point in entering into a conversation (see also Chapter 4).

3.3 Characteristics of Science Communication Processes

In this section, the insights of the communication process models are applied to the field of science communication. Three main elements of the science communication process will be described: the actors or stakeholders involved, motives and objectives to communicate, and the science and technology content and context.

3.3.1 *The Actor or Stakeholder*

Who are the actors or the stakeholders in science communication processes? Due to the various developments that have taken place in the field of science communication, the perspective on the actors, or the parties involved, has changed over time. Initially, the scientists were regarded as the senders, the general public as the receivers, and the journalists and public relations officers as intermediaries. Nowadays, communication about science and technology is rather viewed as a process of interaction between all kinds of senders and receivers, who often change roles during the process. Senders and receivers can be composed of numerous individuals and groups and can be organized in networks of changing compositions, or otherwise. Therefore, each situation has its own relevant parties and in effect its own *public*. This is why the field of science communication, nowadays, often refers to publics in plural, rather than to the public in singular.

Depending on the circumstances, senders can have different names: actors, science communicators, practitioners, communication professionals, scientists, and stakeholders. Receivers also can be referred to as target audience or target group, priority group, visitors, customers, stakeholders, (the general) public, publics, or participants.

In order to be able to easily identify actors involved in a science communication process, they will often be classified in larger groups, such as academia, the government, the industry, the media, civil society organizations (NGOs), and the general public. According to Siune *et al.* (2009), stakeholders are the people or organizations with an interest in a project (see also Figure 3.3). They regard universities and schools as separate stakeholders, in addition to private organizations such as industry and businesses; government and parliament; media; museums; and the so-called third sector. This third sector consists of NGOs, churches, and labor unions, who represent groups of citizens.

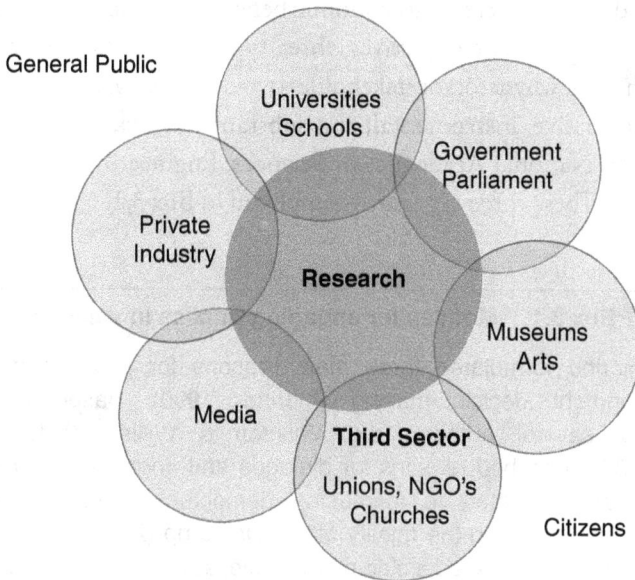

Source: Based on Siune *et al.* (2009).

Figure 3.3: Stakeholders in science communication.

On the outside, the general public, in their role as citizens, has a place. The authors emphasize that: 'In principle, every person in society is a stakeholder when it comes to the role of science in society,' but the 'problem with this concept is that not all of these actors are active' (Siune *et al.*, 2009, p. 20).

Irrespective of which classification will be used, each will involve a simplified representation of groups of people. In practice, communication takes place at many different levels within and between groups, for example, at the personal level, at the departmental level, or at the organizational level. Therefore, each communication situation will differ.

3.3.2 *Diversity of Goals*

All actors in a science communication process work on the basis of certain motives and have particular ambitions or objectives they want to achieve. Sometimes these motives are societal, sometimes organizational or personal.

Science communication is often associated with an increasing involvement of the public in science and technology to achieve societal objectives, such as increasing democracy, creating economic benefits, and increasing the quality of life. Looking at these objectives, three types of societal motives can be distinguished for why actors or stakeholders would want to communicate about science: normative, instrumental, and substantive (Jackson, Barbagallo & Haste, 2005; National Academies of Sciences, Engineering, and Medicine, NAS, 2017). These three types are exemplified in Box 3.1.

Box 3.1: Motives for engaging publics in science

Daniel Fiorino formulated three main reasons for public participation in environment decision-making (Fiorino, 1990). Based on these rationales, several authors (e.g. Wilsdon & Willis, 2004; Jackson *et al.*, 2005) described reasons for dialogue and engagement in science communication, related to increasing democracy, creating economic benefits, and improving the quality of life. Dalderup (2000), related to the latter reason, distinguished a cultural rationale; science and technology are everywhere and are part of people's daily life.

Box 3.1: (*Continued*)

Normative view: Increasing democracy
From a normative view, it is argued that making science more accessible for people enables them to contribute to public discussion about science and technology developments. People have a 'right to know' in current society (cf., Dalderup, 2000; Wilsdon & Willis, 2004; Jackson, Barbagallo & Haste, 2005).

Instrumental view: Creating economic benefits
In an instrumental view, engaging people in science will build greater public trust and confidence in the regulation of science and in scientific institutions. As Jackson, Barbagallo & Haste (2005) pointed out, consultation and greater transparency about decision-making on new developments and acknowledging public interests and concerns will help to reduce the conflict between the scientific community, the regulators, and various publics. It will build trust and therefore lead to economic benefits.

Substantive view: Improving quality of life
From a substantive view, science and technology are embedded in virtually every aspect of modern life. For this reason, people increasingly face the need to integrate information from science with their personal values and other considerations as they make important life decisions, such as those about medical care, the safety of foods, and a changing climate. Therefore, it is important for people to not only know and learn about science and technology but also be engaged in decision-making on how to develop science and technology in a way that benefits society. In this view, it is recognized that knowledge and experience from outside the boundaries of 'academic' science (Jackson, Barbagallo & Haste, 2005) are relevant to improve people's quality of life.

In addition to societal motives, organizations within the science and technology domain may aim at achieving organizational ambitions or objectives. Universities and other research institutions, for example, are organizations that want to work on their reputation, find investors, i.e. sponsors for research, or a sufficient number of customers, in a similar way as businesses do. For industry, it is important to market new technological products, and the media also uses science communication to engage certain target groups. Science communication can even be used to achieve marketing objectives at a personal level. For example, it can help a scientist gain more recognition and therefore obtain funding for research more easily.

Science communication practitioners, ultimately, try to achieve these ambitions by means of communication. In professional science communication, they develop communication strategies. On the basis of a set of coherent choices, they decide on all aspects of the communication process to make this process as effective as possible.

3.3.3 *Science and technology content and context*

Communicating about science can be complicated. As mentioned before, its complexity stems from many diverse and interconnected aspects of the communication processes, including the individuals and organizations involved, their different goals, and their varying interests. However, the nature of information or the state of science itself can also pose a challenge for communication (NAS, 2017).

Science is expected to yield information that is useful to society, whether directly or indirectly. Indeed, science offers a unique set of methods and regulations for producing reliable knowledge about the world. However, scientific findings often represent work in progress or are applicable only to particular contexts or populations. The results of science often provide ambiguous information about questions for which the public wants clear answers. Results can be insufficient or uncertain and scientific conclusions can change over time as new findings emerge (NAS, 2017).

Most people lack familiarity with science in general or with the scientific findings and issues related to a particular decision. People may often feel insecure by the ambiguity of science and tend to have difficulty with understanding scientific uncertainty and probability. In attempts of making sense of complex scientific information, they often use shortcuts based on their beliefs and values: in processing the new information people may rely, for example, on a quick assessment of whether the information fits with what they already know and believe about the subject (Kahneman, 2011). This can lead to inaccuracies in their interpretation of the scientific information, especially when uncertainty is involved (NAS, 2017).

As with other types of communication, trust is an important aspect of achieving communication goals. An audience decides whether the sources of information or the institutions they represent are trustworthy. People use this assessment in deciding what information is worth their attention and often what they think about that information.

In addition to trust, many other factors can influence the acceptance of scientific information. People's opinions and decisions are continuously affected by a variety of social influences, such as social networks, norms, group membership, and loyalties. Therefore, people approach science communication from their own starting points, which are a combination of their expectations, knowledge, skills, beliefs, and values that are in turn shaped by broader social, political, and economic influences (NAS, 2017).

3.4 Developing Communication Strategies

A strategy can form the starting point for developing and aligning communication activities. In order to arrive at a well-founded and coherent strategy, practitioners have to make a lot of coherent choices, such as, for instance, how to align the communication within an organization. An effective communication strategy focuses all communication activities and ensures consistency in communication. Without a communication strategy, it is difficult for a communication practitioner to determine what communication activities are relevant and what are not. Because the environment in which a strategy is implemented is always subject to change, a communication strategy will have to be constantly adapted. The eight interdependent building blocks of the *Strategic Communication Frame* developed by Van Ruler & Körver (2018) could be used for this. Questions raised in the building blocks may help to develop a science communication strategy. The authors of this chapter have adjusted the description of the blocks to the science communication context as follows:

a. **Ambition:** In building a communication strategy, a relevant question for practitioners is: 'What do you want to achieve?' In a dynamic communication context, it is difficult to control the communication process or achieve predefined smart goals. In that case, it is good to think of one's ambition, for instance, to introduce a new innovation, to start a dialogue with different stakeholders about a new technology, or to implement a campaign to prevent obesity. Van Ruler & Körver (2018) define ambition as 'a strong desire to do or achieve something'. So, a starting point for a strategy is to describe the ambition.

b. **Vision**: According to Van Ruler & Körver (2018), ambition is influenced by vision; by people's own perception of their profession and its added value. Therefore, it is important to consider one's own role within the organization and the possibilities and limitations that result from this.

c. **Internal situation**: Science communication practitioners often work in a communication department of a university, government, NGO, or an industry. They represent the interests of their employer. For communication practitioners, it is therefore important to be well informed about and aware of the vision and mission of their organization and to know which priorities the organization adheres to. After all, the communication of an organization should be in line with the organization's goals.

Closely related to the vision and mission of the organization and the ambitions of the communication department are ethical considerations regarding science communication. The decision to communicate science often involves an ethical component. Choices about what scientific results to communicate, when, how, and to whom are a reflection of values that the organization holds (NAS, 2017). For example, it can be questioned whether it is appropriate for an organization to communicate science in order to persuade people to support a particular policy option. Another questionable issue can be to use scientific information, in particular, to encourage people to change behavior, for instance, in health campaigns (see also Chapter 8). In all cases, communication practitioners have to be aware of both the code of conduct of the company as well as their own ethical restrictions, and they must ensure that these ethical choices will be incorporated into the final strategy.

d. **External situation**: In studying the external situation, trends and developments are taken into account as well as the various perspectives that can surround an issue. These trends and developments can play a major role in making choices regarding the implementation of the strategy. An analysis of these aspects is recommended.

Social trends and developments may affect an organization. Distrust in artificial intelligence, for instance, can influence the attitude of the public and their acceptance of a particular innovation in that area. Trends and developments from the perspective of the content of communication can also be relevant. The way in which innovations are framed in the media, for instance, can have a major influence on the way the innovation is perceived by the public (Davies & Horst, 2016). Expectations of the stakeholders or customers are another important issue. If those expectations are clear, the communication can be adapted to it.

All analyses of trends and developments can help communicators to identify what information people need. Besides, the analyses provide them with a realistic view of the situation since communicators 'tend to overestimate what most people know about a subject [...] as well as to overrate the effectiveness of their efforts [...]' (NAS, 2017, p.12).

e. **Accountability:** Van Ruler & Körver (2018) stress that it is important for communication practitioners to make explicit what their precise responsibilities are regarding the stated ambitions and how to measure progress. They have to be able to prove to be on track.

f. **Stakeholders:** In many cases, input of other parties is required to execute a strategy. It is important to make a distinction between enablers and partners (Van Ruler & Körver, 2018). Enablers are the ones whose support the communication practitioners' need to put a strategy in practice. Enablers can be sponsors or *influencers*; individuals or organizations that can help or counteract a strategy. Partners are the people the communicators work with in order to achieve the goal. For each strategy, it is important to consider who could take these roles and how to involve these stakeholders best. Note that all (potential) stakeholders make their own internal and external analysis as well, and therefore place additional demands on the final strategy.

g. **Resources:** Money and people also play a major role in the execution of communication strategies. That is why it is necessary to make an inventory of the required budget and manpower to reach the ambition. It is important to determine whether sufficient knowledge, skills, and experience are available within the team and who will be in charge of putting a strategy into practice.

h. **Approach:** A final step of strategy development is taking into account all requirements from previous blocks and making explicit choices regarding goals, stakeholders, and communication means. In this stage, it is important to prioritize: to decide what needs to be done first and what can be done later, and to make a global plan of activities. (For an overview of communication means, see Box 3.2.)

Box 3.2: Various communication means and their pros and cons

Based on analysis, science communicators ultimately determine which communication means are best to use in their communication strategy. Countless communication means are available, varying from presentations, journals, radio and television programs, websites, social media, science cafés, exhibitions, open days, workshops, panels, science theaters, focus group sessions, and public debates and dialogues.

Some means are more suitable for transmission (one-way communication) and others more suitable for transaction (two-way communication), some can be used for both. For example, science cafés are gatherings where scientists talk about their research in an informal setting and the public has an opportunity to ask questions. These science cafés can be viewed as a means of informing participants and making them aware of developments in science and technology. However, it is also possible for scientists and visitors to exchange ideas, for instance, during the discussion part. Then, a science café is no longer just an instrument for sending information to the participants, but facilitates, at the same time, a process of transaction (Dijkstra & Critchley, 2016).

Likewise, most social media are also versatile communication means, where both transmission and transaction processes are possible. And even during lectures, which are generally regarded as a more transmission-based form of communication, reactions from the audience can lead to lively discussions, which make the communication process far more interactive.

However, choosing a communication means, or rather a combination of means, is not easy. Many possibilities exist, each with its own advantages and disadvantages. Participatory means of communication, for example, such as dialogues or focus group sessions, appear to be a good way of engaging people in a specific subject. On the contrary, only a limited

---

Box 3.2: (*Continued*)

number of people can take part in participatory sessions, the sessions
are relatively time consuming, and not everyone will or can be motivated
to participate actively. Furthermore, designing and facilitating a dialogue
is no easy task. After all, facilitators have to ensure that all participants, as
equal discussion partners, will get the opportunity to contribute their ideas
and thoughts in the limited time available (see also Chapter 4).

Other means of communications have limitations as well. Basically,
mass media make it possible for the information to reach many people
simultaneously. However, traditional mass media such as radio and
television are expensive. Using these resources does not necessarily mean
that the desired target group will be reached, nor that that the communication
has the intended effect. After all, that depends on many factors, such as
whether the message is considered attractive and sufficiently clear.

Social media also have the advantage that large groups of people can
be reached in a short period of time. However, people may or may not
respond to messages and participants may or may not be able to change
or forward messages. Consequently, social media are difficult to control.
Therefore, communication practitioners often do not know who picks up the
message and what the outcome will be.

Using these building blocks does not prescribe what one should do,
or which strategy is best. It helps a practitioner to make grounded choices.
After all, in the dynamic context of science communication, a strategy must
be sufficiently flexible. This requires that during the implementation of a
strategy the answers to the questions raised above must be checked again and
again, and — if necessary — adjusted. Constant reflection and evaluation are
therefore important.

3.5 Using Theories in Science Communication Processes

Theories can be used to better understand a communication process. They
may, for instance, provide insight into the important elements involved
in the communication process and the relationships between them.
Communication theories can therefore be useful in analyzing the current
situation and solving specific communication problems. Theories can also

help to make choices in developing new communication strategies. Besides theory, personal experience and intuition play an important role in the choices professionals make. So, in order to make good decisions, both practical and theoretical insights into how communication processes work are helpful.

Littlejohn, Foss & Oetzel (2017) describe communication theories as, 'any organized set of concepts, explanations and principles that depicts some aspect of human experience' (Littlejohn, Foss & Oetzel, 2017, p. 7). Every theory looks at the process of communication from a different angle, dependent on a different basic belief or perspective on communication.

According to Littlejohn, Foss & Oetzel (2017), most theories consist of four dimensions:

- assumptions or basic beliefs that underlie the theory;
- concepts that are the building blocks of the theory;
- explanations for the phenomenon studied;
- principles that are guidelines for actions.

Various frameworks are available by which communication theories can be arranged. Littlejohn, Foss & Oetzel (2017) describe several of these frameworks. In the Graig framework (Graig, 1999), for example, seven traditions in communication theory are distinguished. All have a different perspective on communication: the semiotic, the phenomenological, the cybernetic, the socio-psychological, the socio-cultural, the critical, and the rhetorical traditions. In Box 3.3, the traditions are briefly exemplified.

Box 3.3: The Graig framework — seven traditions in communication theory (Graig, 1999)

Theories in the *semiotic tradition* focus on signs and symbols. People can give meaning to signs and symbols through their own experiences and perceptions. Because of this, signs and symbols can represent feelings, situations, ideas, or states. Semiotics has been especially important in helping to understand messages. It does not focus on the people involved in a communication process.

Box 3.3: (*Continued*)

Theories in the *phenomenological tradition* assume that people actively interpret their experiences and come to understand the world by means of their personal experiences and values. While in the semiotics, interpretation is considered to be separate from reality; in phenomenology, interpretation literally forms what is real for the person.

The *cybernetic tradition* examines the overall workings of communication in systems, in which elements interact and are dependent on other parts. To use an example, in a classroom system, the relationships between the students and teacher and among students themselves, subject matter, environment of the classroom, cultural diversity of students, and homework all come together to form a cycle of networks and connections. As Littlejohn, Foss & Oetzel (2017, p. 41) put it: 'Systems monitor, regulate and control their outputs in order to remain stable and to achieve goals'.

The *socio-psychological tradition* focuses on the individual. Much of the work in this tradition has focused on how people develop and process messages and on the effect of messages on individuals. Several theories relate to change of attitude or behavior of people by means of communication. Much of the work in risk communication (Chapter 7), health communication (Chapter 8), and environmental communication (Chapter 9) is associated with this tradition.

Researchers in the *socio-cultural tradition* want to understand ways in which people *together* create the realities of their social groups, organizations, and cultures. Theories within this tradition focus on patterns of interaction between people.

Theories within the *critical tradition* are concerned with how power, oppression, and privilege are the products of certain forms of communication throughout society. In the field of communication, critical scholars are particularly interested in how messages reinforce oppression in society.

Finally, rhetorical theory developed over time. Central to the classical *rhetorical tradition* was the art of persuasion. These theories originated when people began to identify that listeners could be influenced by speeches. They focused on how the element of persuasion works and how it can be utilized effectively. Nowadays, an important notion in rhetorical theory is that humans create their worlds through symbols: the world they know is offered to them by language.

3.6 How Communication Theory Informs Communication Practice

To illustrate the added value of using communication theories in communication strategies, three short cases will be discussed, each with a focus on how and to what extent communication theories can be used. Too detailed information about the setting of the case has been omitted to ensure focus on the purpose of this section.

Case 1: Communicators want to persuade their audiences
Science communicators want to make sure that their target audience is susceptible to the message. How to achieve that? How to persuade the audience to process the message, especially considering the fact that today everyone is overloaded with hundreds of messages per day?

Theory: A communication theory that relates to information processing is the Elaboration Likelihood Theory (Petty & Cacioppo, 1986); a theory in the socio-psychological tradition. The theory describes two ways for processing persuasive messages: the central route and the peripheral route. When people process information through a peripheral route, they are basically not very critical to the information, but quickly make judgments based on a few cues. Information processed through the central route, however, makes sure that people process the information more intensely and assess the information more actively with regard to what they already know. Processing information centrally will more likely lead to a lasting attitude change.

To take the central route, people have to be able to elaborate on the argumentation that is provided. They have to be motivated to process the message and to be able to do that with regard to the availability of time, opportunity, and cognitive abilities. In practice, these criteria are often not met. Then, the peripheral route remains. To make people take the peripheral route, a communicator can use several means: give arguments; make sure that the information relates or partially overlaps with something a person already knows; make sure that the source of the communication message is seen as credible by the recipient, for instance, via an influencer or by making sure that the communicator is a well-respected party; and ensure an attractive packaging of the message.

Practice: Infotainment — the combination of information and entertainment — is both an internationally and widely used method to reach large numbers of people and specific target groups. Soaps use the peripheral route to influence the attitude of the general public, for example, with respect to the acceptance of LGBT people, education about a healthy lifestyle, and sustainable behavior.

Case 2: The strength of identification

In professional life, many people make use of speeches to create bonding with the audience. But, what makes a good speech? Why do some speeches really create a sense of belonging while others do not?

Theory: The theory of identification of Burke (as explained by Littlejohn, Foss & Oetzel, 2017) is a theory in the rhetoric tradition. It states that language is by nature selective and abstract. It sets certain parts of reality in the spotlight and ignores others. Furthermore, language contributes to human action by bringing people together or dividing them. When the symbols used in a language bring people together, a common way of understanding, that is, identification, occurs. However, the opposite, division, may also happen. A language can promote separation and division by emphasizing differences.

Burke distinguishes the following four kinds of overlapping sources of identification which bring people together:

- material identification (e.g. owning the same goods);
- idealistic identification (e.g. having similar values, interests, experiences);
- formal identification (e.g. based on the identification that arises from the form, arrangement, or organization in which both parties participate);
- and identification through mystification (e.g. people at the lower strata in a hierarchy often identify with people at the top hierarchy).

If communicators want their audience to identify with the speaker or with the speech content, they could adjust the speech to the right source of identification. In this way, they can convey messages that lead the audience into entering very different kinds of worlds. However, a disadvantage is that the theory can only be applied in face-to-face communication. Besides that, if the attempt to achieve identification is exaggerated, the strategy could backfire.

Practice: A famous example of the use of idealistic identification is 'Ich bin ein Berliner' ('I am a Berliner'), a quote from the speech by the American President John F. Kennedy during the Cold War in West Berlin in 1963. In this quote, he used the identification with the Berliners to underline the support of the United States to West Berlin. The speech was experienced as a moral boost for the West Berliners, who lived in an enclave in the German Democratic Republic and feared a possible Soviet occupation.

Case 3: Group behavior
In a lot of communication settings, groups are involved. Take, for instance, a research team or a department of an organization. In most cases, a desire for harmony or conformity exists in the group. Group members or managers often try to minimize conflict. However, this can also have disadvantages.

Theory: The work of Irving Janis and his colleagues has been influential within group communication literature. His *Groupthink* theory (Janis, 1982) explains how groups interact toward decision-making depending on the degree of interdependence in a group. Like most theories that relate to group structure and group tasks, this is a theory in the socio-cultural tradition. Members of a cohesive group rely on one another to achieve certain mutually desired goals. The more cohesive a group, the more pressure it exerts on the members to maintain that cohesiveness, and the more likely *Groupthink* will happen. *Groupthink* requires individuals to avoid raising controversial issues or alternative solutions, and loss of individual creativity, uniqueness, and independent thinking occurs. The dysfunctional group dynamics of the so-called *in-group* produces an 'illusion of invulnerability' (an inflated certainty that the right decision has been made). Thus, the in-group significantly overrates its own abilities in decision-making and significantly underrates the abilities of its opponents (the 'outgroup').

Groupthink can cause closed mindedness, can limit the amount of alternatives the group takes into account, and can prevent ideas from being examined critically. It can also make the group so confident of its own ideas that it does not consider contingency plans. Furthermore, *Groupthink* can produce dehumanizing actions against the 'outgroup'.

Janis provides advice on how to prevent or solve the problem of *Groupthink*. For instance, do not have the leader state a preference up front, invite outsiders to bring in fresh ideas, and assign someone to be the devil's advocate (Littlejohn, Foss & Oetzel, 2017).

Practice: In schools, the *Groupthink* process is often clearly visible. *Groupthink* also occurs in politics and in business, as in many situations and systems 'wanting to belong to the group' is of great importance.

From the cases described above, it becomes clear that the added value of communication theories sometimes can be very concrete. For instance, the tips of Janis (1982) to avoid *Groupthink* are directly applicable. Mostly, however, theories give communicators insights which are only relevant in specific situations. This applies, for instance, to Burke's theory of identification. For each and every speech, one needs to determine what sources of identification are relevant and useful for this particular topic, context, and audience. Therefore, theories often give guidance and insight, but not immediate concrete answers. For science communicators, it is advisable to take a step back and reflect on which theory would fit the current communication situation and produce the desired outcome in a best possible manner.

3.7 Conclusion

Communication can be complicated. In professional communication, various aspects have to be taken into account to achieve particular communication goals or ambitions. Communication models can help to point out the most relevant aspects and their mutual relationship. It is important to note that each model has its own focus.

Science communication processes have some particular characteristics and challenges, related to the content and context of science and technology, the stakeholders involved, and their different motives and interests.

In organizations, communication activities have to be aligned. A strategic communication framework can help science communication practitioners to develop effective communication activities in an organizational context. The framework can be built by making explicit and coherent choices with regard to different themes (building blocks). A final step in building a communication strategy is to translate these choices in the various building blocks into a set of concrete communication activities, to prioritize activities, and to plan them. Using these building blocks does not prescribe what one should do, or which strategy is best. It helps a practitioner make substantiated choices. Because of the dynamic context of science communication, constant reflection on and evaluation of the communication activities and the framework is therefore of importance.

A theory can play an important role in developing communication strategies. It can provide insight into communication processes, can be useful in analyzing the current situation, and can provide points of reference when all strategic decisions have to be translated into operational communication activities. However, communicators should ask themselves for each communication task what theories will fit the task at hand best.

References

Davies, S. R., & Horst, M. (2016). *Science Communication, Culture, Identity and Citizenship*. London: Palgrave Macmillan.

Dalderup, L. (2000). Wetenschapsvoorlichting en wetenschapsbeleid in Nederland 1950–2000. *Gewina: Tijdschrift voor de Geschiedenis der Geneeskunde, Natuurwetenschappen, Wiskunde en Techniek, 23*(3), 165–192.

Dijkstra, A. M., & Critchley, C. R. (2016). Nanotechnology in Dutch science cafés: Public risk perceptions contextualised. *Public Understanding of Science, 25*(1), 71–87. doi: 10.1177/0963662514528080.

Fiorino, D. J. (1990). Citizen participation and environmental risk: A survey of institutional mechanisms. *Science Technology & Human Values, 15*(2), 226–243.

Graig, R. T. (1999). Communication theory as a field. *Communication Theory, 9*(2), 119–161.

Jackson, R., Barbagallo, F., & Haste, H. (2005). Strengths of public dialogue on science-related issues. *Critical Review of International Social and Political Philosophy, 8*(3), 349–358. doi: 10.1080/13698230500187227.

Janis, I. L. (1982). *Victims of Groupthink: A Psychological Study of Foreign-Policy Decisions and Fiascoes*. Boston: Houghton Mifflin Co.

Kahneman, D. (2011). *Thinking, Fast and Slow*. New York: Macmillan Publishers.

Lasswell, H. (1948). The structure and function of communication in society. In L. Bryson (ed.), *The Communication of Ideas*. New York: Harper and Brothers.

Littlejohn, S. W., Foss, K. A., & Oetzel, J. G. (2017). *Theories of Human Communication*. Long Grove, Illinois: Waveland Press, Inc.

National Academies of Sciences, Engineering, and Medicine (NAS). (2017). *Communicating Science Effectively. A Research Agenda*. National Academy of Sciences: Washington, DC. Retrieved March 26, 2019 from https://www.nap.edu/catalog/23674/communicating-science-effectively-a-research-agenda.

Oomkes, F. R. (2013). *Communicatieleer: een inleiding (Communication Studies: An Introduction)*. Den Haag: Boom Lemma.

Petty, R. E., & Cacioppo, J. T. (1986). *Communication and Persuasion: Central and Peripheral Routes to Attitude Change*. New York: Springer.

Siune, K., Markus, E., Calloni, M., Felt, U., Gorski, A., Grunwald, A., Rip, A., de Semir, V., & Wyatt, S., (2009). *Challenging Futures of Science in Society, Emerging Trends and Cutting-edge Issues*. Report of the MASIS Expert Group setup by the European Commission. Brussels: European Commission. Retrieved March 26, 2019 from http://www.securepart.eu/download/com-2009_masis_report_expert-group-_en150625092421.pdf.

Van Ruler, B., & Körver, F. (2018). *The communication Strategy Handbook: Toolkit for Creating a Winning Strategy*. New York, NY: Peter Lang Publishing USA.

Wilsdon, J., & Willis, R. (2004). *See-through Science. Why Public Engagement Needs to Move Upstream*. London: Demos. Retrieved March 26, 2019 from https://www.demos.co.uk/files/Seethroughsciencefinal.pdf.

Chapter 4

Science in Dialogue

Roald Verhoeff and Frank Kupper

4.1 Introduction

Contemporary society is confronted with complex issues — climate change, the increasing scarcity of raw materials and nutrients, and the aging population, among others. Scientists play an important role in identifying, analyzing, and finding solutions to such issues, and this role has changed significantly over the years. As shown in Section 2.8 of Chapter 2, scientists are not only asked to provide reliable knowledge that businesses and government agencies can apply but also increasingly requested to contribute to social issues through providing socially robust knowledge that benefits health care, the economic power of individual countries, or governmental sustainability policies. To develop socially robust knowledge, scientists are working more and more with other stakeholders in society such as local and national governments, businesses, non-governmental agencies (NGOs), experiential experts, or other members of the general public.

This chapter focuses on the changing role of science in tackling complex social issues. Two trends are discerned with respect to science in dialogue: (1) research into the ethical and social aspects of science and (2) public participation in science in which interaction between social parties or experience experts and scholars is central. Within the first trend, the term *dialogue* refers to the interdisciplinary collaboration between natural scientists, social scientists, and philosophers in opening up the social practice of science to society. Within the context of *public participation*, dialogue refers to the reciprocal nature of communication processes in which public stakeholders can contribute to decision-making about developments in science. Communication professionals regard dialogue as a method for shaping public participation. At the end of this chapter, some characteristics of dialogue in practice and the diversity of roles of a communication professional in the dialogue process are presented.

4.2 Controversies and Complex Issues

In opening up science to society, the increasing interest in public participation in science is related to the broader social trend of democratization that started in the 1960s. This has gone hand in hand with people becoming more educated as a result of a major expansion of higher education. In addition, besides the increasing prosperity that has accompanied technological advances, the adverse effects of new technologies have become more visible: the effects of the atomic bomb in Hiroshima, climate change, or the nuclear disaster in Chernobyl, for example. Such events made it less obvious that science should develop in a relatively autonomous way (Bauer, Allum & Miller 2007; see also Chapter 1). Nowadays, the newspapers are full of articles and stories on controversies and complex issues, and consensus in science seems to be absent: How to tackle cybercrime? Should engineers proceed with shale gas drilling? How to deal with climate change and related problems, such as the availability of water in dry areas? How can local officials protect people from risks such as hurricanes, earthquakes, floods, or volcanic eruptions? What can be done about food waste in Western countries while in remote and poor areas people are starving? Science is involved in these issues in two ways.

Science offers solutions to social problems but not without creating new problems or risks. For example, food production has become much more efficient thanks to science, but the development of agricultural technologies has gone hand in hand with the increasing depletion of available resources and the incidence of food crises such as the BSE epidemic in the 1990s. Another example is technology-induced human enhancement. Technical intervention allows science to offer increasing opportunities not only to help those suffering from disease or illness but also to improve the functioning of healthy people. For example, plastic surgery, performance enhancers like steroids, or fertility treatments are all meant to make mankind attractive, smarter, and more (re)productive. Since the 1970s, an increasing awareness with regard to social implications of science has come about, for instance, potential health risks, the distribution of wealth, access to health care, and basic questions about human identity.

This illustrates the complexity of policy issues that arises at the interface of science and society. In the social sciences, these issues are referred to as unstructured problems or *ill-defined wicked problems* (Hisschemöller & Hoppe, 1996, Hisschemöller *et al.*, 2001; Rittel & Webber, 1973; see also Figure 4.1). These problems are difficult to define because they are characterized by both a high level of technical uncertainty and a high degree of normative diversity.

Technical uncertainty refers to the fact that science cannot provide ready-made solutions. This may be because the problem has not been fully defined yet or because insufficient knowledge is available for finding a solution. In addition, scientists can draw different conclusions from the research that has already been done, resulting in a lack of consensus on the right approach.

Pluralism of values, interests, and viewpoints
High   Low

		High	Low
Certainty of relevant knowledge	No	Wicked problem (e.g. climate change, pandemic influenza, disposing nuclear waste)	Moderately structured problem (discussion about suitable meansto deal with the problem, e.g. traffic jams)
	Yes	Moderately structured problem (discussion about ethical acceptability of goals; e.g. pre-implantation genetic testing)	Structured problem (CFK gasses affecting the Ozonelayer; fixing a car)

Figure 4.1: Four types of policy problems: From structured problems in which science can or has already provided solutions to wicked problems in which science cannot offer ready-made answers.

Finally, the (social or natural) system under study, such as the climate system, may be so complex that some degree of uncertainty is inevitable.

Apart from these technical uncertainties, normative diversity also exists. People have different interests and ideas on how to prioritize and tackle contemporary problems. In line with their beliefs, ideals, experience, or political orientation, they all bring their own perspectives to social issues and possible solutions. Is it a good idea to reward a sustainable lifestyle with governmental subsidies, or can policy-makers achieve better results by raising fuel taxes? Should the government deal with climate change by investing in relatively expensive storage of CO_2, or should it reduce CO_2 emissions by raising taxes on energy use? Or perhaps, it is best to do both. In these questions, it is difficult or perhaps impossible to draw a line between science and politics.

The complexity of today's problems is increasingly recognized by various stakeholders, i.e. governments, NGOs, industry, and science. This recognition is a result of the contextualization of science (see also Chapter 2), in which science functions more and more as a 'recognizer of problems' or a 'mediator'. This means that knowledge is increasingly being produced in collaboration with other social actors such as businesses that are actively involved in research on renewable energy or patients who are involved in shaping biomedical research agendas. Science that includes relevant social values and perspectives in analyzing and solving social problems has been called 'socially robust science' (Gibbons, 1996, 1999).

Since the 1990s, the trend to include social values and perspectives in science has become visible in two parallel movements: ethical and social research, and public participation. Public participation shapes the involvement of various (future) stakeholders in the reflection on new knowledge and technology: science in dialogue with society. Both public participation and ethical and social research focus on the social embedding of scientific research and the opening up of science to societal demands.

Socially robust knowledge includes both the scientific knowledge of the natural scientist as well as the professional and lifeworld knowledge of social stakeholders and experiential experts. Experiential expertise, for example, could refer to farmers in issues around livestock breeding or the use of GM crops. Lifeworld knowledge could refer to the experience of patients after having a medical treatment, for example. The ethical and social research on

Table 4.1: Examples of ethical and social aspects of new technologies.

Technology	Innovation/application	Ethical and social aspects
Biobanks	Storage of human tissue for clinical and research purposes	Intellectual property issues, privacy
DNA test	Test for hereditary breast cancer (genes BRCA1 and BRCA2)	Intellectual property issues (patenting of the DNA test), privacy
Optimization of food crops	Use of genetically modified seeds	Intellectual property issues such as patents of multinationals vs local farming culture and communities, environmental effects (biodiversity)
Biofuels	Genetically modified yeast cells or algae that convert (non)edible parts of plants like wood, straw, or waste into bioethanol	Use of scarce agricultural land for fuel instead of food, release of fine particles when burned, sustainability
Nanotechnology	Use of nanoparticles in consumer products like sunscreen, sports drinks, paint, clothing, and adhesive tape	Unintended health and environmental effects

science and technology, known in the 1990s by the acronym Ethical, Legal and Social Aspects of Science (ELSA), focuses on mapping the ethical and social implications of science as exemplified in Table 4.1. These studies are based on disciplines such as philosophy, bioethics, and sociology. This is where scientists are in dialogue with other social scientific, philosophical, and ethical experts.

4.3 Reframing Science in Society

In the 1970s, interest in the social sciences increased significantly and also resulted in an increased attention for research on the ethical and social aspects of science and technology. In ethical and social studies, ethicists and social scientists aim to elucidate potential social implications in collaboration with natural scientists at an early stage of developing new knowledge or technologies. This kind of research received a major boost when the director of the Human Genome Project (HUGO) in 1988 announced that HUGO would have many ethical and social implications. It was decided to devote a significant portion of the financial resources of HUGO to research into these implications. This trend in research funding has been followed by many countries, including the Netherlands. In 2002, the Dutch government initiated the Netherlands Genomics Initiative, which spent 5% of its budget on ethical and social research during its existence until 2013.

Table 4.1 shows some examples of ethical and social issues concerning new technologies. With scientific developments that are still at an early stage, it is difficult to properly establish and assess the exact circumstances in which the technology will be used. No one, for example, could predict the widespread use of the Internet when it was developed in the early 1970s as a service of the US Department of Defense. Another example is the unexpected use of Viagra as a drug stimulating sexual performance, while it was originally developed as a medicine for improving the blood flow in the heart muscle.

Exploration of ethical and social aspects in an early phase of development is important for the effective redirection of technological innovations toward social entrenchment while it is still possible. This could increase their implementation in practice by having an impact on how the technology is presented, how data are stored, how rules or laws are amended, as well as, ideally, on the reorientation of research programs themselves. It is not only important to look at the direct (intended) applications and users but also at the potential unintended uses and implications (Gibbons, 1999; Von Schomberg, 2007) and who can or should be responsible for them. This is connected, for example, to whether a technological innovation sufficiently contributes to (or conflicts with) the quest for social justice and sustainability. Researchers into ethical and social aspects of science hold a position similar to that of experts in science communication: they are close enough to the science to be able to signal new developments but have enough distance to be able to reflect on these developments from a broader perspective or to raise and organize discussions on science and technology.

In order to assess societal implications of future developments and steer science and technology advancement toward acceptable ends, scientific expertise alone is not sufficient anymore. The old 'contract' between science and society, in which science shared its reliable knowledge with society in exchange for autonomy, is gradually being replaced by a new reality (Gibbons, 1996). Science is increasingly expected to consider the ideas, needs, and concerns of consumers, patients, and citizens in the production of knowledge. Society now speaks back to science. Moreover, in the past decades it has witnessed a myriad of ways in which non-scientific actors, such as governments, companies, societal organizations, and citizens, contributed actively to the production of knowledge. In other words, the boundaries between science, government, industry, and citizens have been blurred and are increasingly crossed back and forth by a variety of actors.

This exchange requires various interaction and communication processes, as is reflected in the research policy of the European Union: Horizon 2020. This program has set the agenda for research funding from 2014 till 2020. It uses the concept of Responsible Research & Innovation, which aims at doing science with and for the society. The European Union supports the role of society in research as is also reflected by the 'RISE High-Level Group'. This is a group of Research, Innovation and Science Policy Experts who stress openness and diversity in research and innovation in shaping Europe's future, *Open science, Open Innovation, Open to the World* (Directorate General for Research and Innovation, 2018). This document refers to a research and innovation approach in which partners from the industry, (local) governments, and NGOs are involved at an early stage of development in acquiring relevant knowledge about experiences, social values, and priorities. They also look at the acceptability of different options for designing more sustainable products and applications. Mutual learning of stakeholders who have different expertise, experiences, interests, concerns, and worldviews is key in such collaborations to include as much social perspectives in scientific research as possible.

4.4 Reframing Science–Public Interactions

Research into the ethical and social aspects of science and technology has broadened the basis for decision-making in science by including non-technological disciplines such as sociology and ethics. Public engagement and dialogue go one step further and allow non-experts to influence decisions about the practice and future of scientific research and technological innovations. This extension of the expert forum is an important element in the development of socially robust science (see also Chapter 2).

The increasing role of public dialogue in science is related to the broader social trend of democratization that started in the 1960s. In the 1990s, this trend culminated into a rise of public engagement practices across Europe and the US. Partly, this new mood for dialogue has been due to the perceived need to re-negotiate the legitimacy of science, in response to a series of badly managed public controversies surrounding new technological developments such as nuclear energy and biotechnology. Another important factor has been the increasing recognition of other sources of knowledge and ways of knowing as potentially valuable contributions to policy- and decision-making about science. A pioneering development in this respect has been the involvement of farmers in agricultural innovation and the participation of patients in biomedical research. As a result, the one-sided transmission of information to the public via mass media or education was no longer regarded as sufficient to build trustworthy relationships between science and the public. Science communication theory and practice shifted toward models of dialogue and participation, albeit that the deficit model of communication never disappeared and often remained dominant (Brossard & Lewenstein, 2010). Nevertheless, this shift has been referred to as the move from deficit to dialogue (see also Chapter 1).

The models of dialogue and participation represent a type of communication that is reciprocal in nature. This reflects the increased recognition of the local knowledge of public communities and their capacity to understand technological information and to make a valuable contribution to policy decisions about scientific developments (Burgess, 2014). The recognition of local or contextualized knowledge requires the involvement of individuals and groups in society who can provide this knowledge, and therefore also the involvement of 'reflexive' scientists who are willing and able to critically reflect on their own expertise and perspective. The role of the communication professional is to facilitate and optimize the interaction between the different participants of dialogue processes and events.

4.5 The Nature of Public Dialogue

4.5.1 Three Motivations for Participation and Dialogue

The organizers of participative processes or dialogue events between science and society often refer to three main reasons for involving different stakeholders (Wilsdon & Willis, 2004). First, the substantive argument: the involvement of social stakeholders in science and technology improves the quality of the decisions taken. Too often, science offers only a limited view on the complexity of the issue. Due to the input of knowledge from other, non-scientific perspectives, a more comprehensive definition of the problem may be achieved and it is possible to generate more socially robust solutions. One example is the involvement of patients in the biomedical research agenda. Patients can talk about their experiences in dealing with their illness in daily life, which can result in attention for issues that are under-researched by science.

The second motivation for the participation of social stakeholders refers to the instrumental argument. The involvement of society at an early stage of scientific development and innovation leads to more support for research results and reduces the risk of public resistance to new innovations. This increases the chance of the successful implementation of new technologies in society.

Finally, the normative argument for participation: citizens have the democratic right to state their views on matters that concern them. Scientific and technological advances usually have a great impact on people's lives and are often financed, at least partly, by public funds. Dialogue is an important means for involving citizens in decision-making regarding scientific and technological developments.

In recent years, public participation and dialogue have become part of an international trend in science policy. By involving public stakeholders in decision-making processes regarding science and technology, policy-makers aim to reduce the perceived lack of credibility. As indicated in the introduction to this chapter, key actors in the policy arena, such as the European Union, have embraced the involvement of civil society. In publications of and grants by national research councils, dialogue is considered a prerequisite for the successful implementation of science in society.

4.5.2 Policy-informing Dialogue and Mutual Learning

Davies *et al.* (2009) have distinguished two types of dialogue events: events that are aimed at informing policy and events that facilitate mutual learning between participants. This distinction has important consequences for the communication professional who wants to organize a dialogue event in which one of these aims is central.

Policy-informing dialogue aims at finding consensus between different stakeholders to formulate a shared advice for policy-makers as exemplified in Box 4.1. For evaluating such dialogues, the actual impact on policy is key. Science communicators organizing such an event should not only ensure that the process is transparent but also provide a formal structure that ensures that the outcome influences the actual process of policy-making. Informing and instructing participants is important for convincing them to bring in their expertise and to contribute to the common goal of formulating shared advice.

Box 4.1:   Policy-informing dialogue for consensus on climate change

Source: Photo credit: Roald Verhoeff, IPCC meeting Stockholm, September 2013.

Box 4.1: (*Continued*)

In the discussion on climate change, uncertainties are omnipresent due to the size and political load of the problem. Science cannot offer any certainty in such complex issues. The eventual sea-level rise depends on human activities — and thus political decisions taken by governments — now and in the near future. The uncertainties in climate change go beyond the technical level and also concern normative considerations: how to weigh the prosperity and interests of countries, different cultures, future generations, and the extinction of species?

Despite this complexity, an international policy on climate should be formulated that takes the uncertainties, interests, and values of all national governments involved into account. The basis for this policy is formed by the Intergovernmental Panel on Climate Change (IPCC, www.ipcc.ch), which includes hundreds of experts from around the world and from universities, research centers, enterprises, environmental, and other organizations. They do not conduct research on behalf of the IPCC but evaluate research published in scientific journals.

The IPCC aims at consensus on the causes and consequences of climate change and looks out for opportunities to adapt to and reduce greenhouse gas emissions (the photo was taken during the IPCC meeting in Stockholm in September 2013). As many as 829 authors from over 80 countries contributed to the Fifth Assessment Report on the physical science basis of climate change published in 2014. Subsequently, at a plenary meeting, government delegations from 198 countries approved and adopted the report, including the executive summary for policy-makers. This summary for policy-makers was discussed phrase by phrase (and amended as necessary), and each delegation had an equal vote.

The outlines of the Sixth Assessment cycle have been approved by the panel at its 46th session in September 2017 and will be finalized in 2021 (www.ipcc.ch). This time 721 experts from 90 countries have been invited to participate in this sixth cycle as authors, reviewers, review editors, and government representatives. The involvement of many experts and governments should ensure the legitimacy of the decision-making process and increase public support. Nevertheless, the IPCC has also been criticized: critics say climate skeptics are not heard enough and the process is cumbersome and not sufficiently transparent to the public.

Non-policy-informing dialogue events or mutual learning events focus on the exchange of ideas and perspectives of the participants. Mutual learning requires a safe environment in which participants can engage in an open conversation as equals. The aim is not to reach an agreement but to generate different ideas and to support the mutual learning process. Public participants learn about new developments in research and technology and their social implications. Scientists develop a better understanding of the social or everyday life context in which the results of their research will be used. By interacting with each other, scientists will also be empowered to shape their social responsibility. Public participants learn to clarify their ideas and can carry out a personal assessment of the risks and benefits of the innovation at an early stage of development. In this way, dialogue contributes to a step-by-step change and clarification on the role of science in society. Box 4.2 presents an example of a non-policy-informing dialogue project.

Box 4.2:   Mutual learning dialogue on synthetic biology

Synthetic biology is a new, emerging field of science in which segments of DNA that are made from scratch or that are taken from different species can be put together to make new DNA structures. In this way, new living biological systems are created with functions that do not exist in nature. This area of research holds a great promise for innovation in fields as diverse as health care and sustainable energy. Simultaneously, it may proliferate risks for humans, animals, and the environment and invoke dilemmas with respect to issues such as equality, autonomy, and identity. Synthetic biology may more and more become a subject of public controversy.

The question for civil society is how to deal with this potential variety of impacts. Synthetic biology does not merely happen to us. It is a technology in the making. Achieving socially robust knowledge and applications of synthetic biology requires reflection and dialogue, aimed at mutual learning about future scenarios, values, and viewpoints among the actors involved as well as the society as a whole. The European Commission has funded a Mobilisation and Mutual Learning Action Plan aimed to further develop the European dialogue, the SYNENERGENE project (www.synenergene.eu). Within the context of this project, a mix of over 20 knowledge institutions, societal organizations, and intermediaries across the EU joined forces to develop reflection tools, dialogue processes, and platforms and to carry out a range of reflection and dialogue activities throughout Europe.

Box 4.2: *(Continued)*

In the 4 years from 2013 to 2016, the SYNENERGENE project generated a lot of public attention for the relevance and implications of the emergence of synthetic biology in society. They organized many different opportunities for reflection and dialogue on the subject, involving a broad range of stakeholders and citizens. The activities that have been developed as part of this project range from (1) the use of arts, film, and theater activities to make participants think and spark dialogue about the social and cultural meanings of synthetic biology to (2) the development of reflection tools and resources for science education and informal learning activities and (3) the development of stakeholder workshops to increase alignment and mutual learning among professional stakeholders, for instance, scientists and policy-makers, involved in the debate.

4.5.3 *Relevant Publics*

Often, even in this chapter, science and the public are referred to in general terms, as if they are both indivisible, monolithic entities. But in reality that is not the case. From academic debate, a sense emerges that the so-called public is a multifaceted collection of different individuals and groups, each with their own ideas, values, assessments, and assumptions. Indeed, even as an individual, one will be torn between conflicting values and ideals when trying to form an opinion about a (socio-)scientific controversy. Due to the complex and ambiguous nature of a wicked problem, the different perspectives on any single problem can be deemed all the more valuable.

The American philosopher and pragmatist, John Dewey argued in his essay *The Public and its Problems* (1927) that the public is not a given unit but is shaped by the issue that is being discussed. The public is not a permanent community of individuals but arises when people are united around a particular issue. An audience emerges when people are affected by a particular social development. Take the example of the introduction of a new vaccine against human papilloma virus, which can cause cervical cancer. Twelve-year-old girls, their parents, doctors and teachers, as well as people who are critical of vaccination for religious reasons have been confronted with this vaccine and have therefore become a public. Note that the members of this group can be strangers to each other: even though they are united around the issue in question, they still differ in many other respects.

Organizers of a social dialogue on science or technology, therefore, need to design processes through which publics can be organized in practical or institutional ways, so that their viewpoints can be represented and highlighted. Dialogue events should therefore have a design with space for the different publics, who are directly or indirectly affected by any given issue, and their views. Publics can include people with a shared religious orientation or youngsters as the generation of the future who will suffer the consequences of climate change.

4.5.4 Decision Space

In recent years, the focus of public engagement and dialogue activities has moved upstream toward early stages of research and innovation trajectories. Early on in the research trajectory, choices still have to be made about the direction of new developments. At that stage, the chance that public perceptions have already solidified into deep-rooted conflicts is small (Wilsdon & Willis, 2004; Felt et al., 2013). These early stages of research and innovation, it is argued, present an excellent opportunity to involve societal actors in making such choices and steer development toward acceptable ends.

Early involvement is not without problems, however, as a great deal of uncertainty about the actual technical and social implications exists. Even experts may not be aware yet of the existence of a technology and therefore will be less inclined to participate in social debates. On the contrary, at a later stage of development the innovation may already be implemented in certain practices. As a result, the only policy options may be to discuss the conditions that have to be met for broader implementation. This can result in public opposition because people feel forced to accept an innovation without having had sufficient input in its development. The tension between involvement at an early and at an advanced stage of innovation has become known as the Collingridge dilemma (Collingridge, 1980).

Public debates on biotechnology and genetically modified crops (GM crops), in particular from 1990 till now in Europe, constitute a notorious example of a controversial technology that, as long as the public debate is managed badly, will continue to result in public outrage. Various NGOs in Europe revolted against GM crops because they had not been involved in decision-making at an early stage. The public resistance that emerged as a consequence resulted in a ban on the import and sale of GM crops in Europe from 1998 to 2004.

When organizing a dialogue at an early stage of technology development, it is helpful to invite experts who will inform participants about the (im)possibilities. Such experts could provide an assessment of the benefits and risks of new technologies and present alternatives. Other experts could elaborate on the ethical or legal implications, personal dilemmas, or socio-political controversies that might arise.

4.5.5 Conditions for Dialogue

When is communication considered to be a dialogue? Confusingly enough, the term dialogue is used both for a meeting between two or more persons as well as for a large-scale social debate and even for the exchange between science and society as two abstract entities. The core of dialogue, however, is the transaction of knowledge that occurs, rather than the transmission of knowledge. Dialogue is about the free exchange of ideas and opinions between two or more parties, institutions, or groups, such as 'doctors and patients' or 'climate scientists and policy-makers'. The goal can be to increase mutual understanding, to find common ground, or to find solutions to social conflicts.

Many ideas about dialogue between science and society are founded on the ideals of the German philosopher Habermas: communicative action in which actors in a society seek to reach common understanding and coordinate activities by reasoned argument, consensus, and cooperation as opposed to strategic action strictly in pursuit of their own goals (Habermas, 1984, p. 86). In the ideal situation, the exchange of arguments is not hindered by power relations. Participants are willing to share and critically reflect on their own assumptions and to adjust them when necessary. According to another German philosopher Gadamer (1960), it is a matter of being open to the other, accepting that someone with a different viewpoint can, at least potentially, make valid arguments.

The Dutch humanist Smaling (2008) proposed the following conditions for what qualifies as a dialogue:

- *Equality*: Every participant is allowed to (partly) define the issue to be discussed, ask questions, and partake in the discussion.
- *Mutual trust and respect*: The participants assume that the dialogue partner is honest. Through their behavior and language they show that they tolerate each other, precisely in 'being different'.
- *Mutual openness and understanding*: Participants contribute relevant knowledge, emotions, and uncertainties and try to interpret and understand each other as much as possible. They support each other in introducing each other's arguments in the best way possible.
- *Argumentative quality*: The participants try to substantiate their reasoning, theories, claims, and opinions on the basis of good or acceptable arguments. Discussion tricks and fallacies, such as the appeal to authority, should be avoided.
- *Reflective and evaluative nature*: The participants reflect on their own communicative role and assess the quality of the dialogue in line with the preceding conditions.

Meeting these conditions increases the collaborative nature of the dialogue in which participants learn from each other and develop ideas together on a shared issue. This contrasts to the one-sided transmission of information in science communication which mainly results in the public gaining new insights. In a dialogue, scientists themselves obtain new insights as well. This collaborative nature results in a fundamentally different communication process than what happens during a debate. In a debate, participants try to win each other over, as can be often seen in politics. One can also determine the winners and losers in debates. A debate provides keen insight into the arguments behind the different perspectives when parties disagree and can force a political decision. A sharp debate may also increase the contrast, making it more difficult to find a solution.

The above conditions for what qualifies as a dialogue are not easily met and demand a great deal from the participants. Therefore, dialogue needs a facilitator or moderator, a role that a communication professional can fulfil. It is the job of the facilitator to create a safe environment in which everyone feels free to share their concerns and ideas. He or she must also entice people, confront them, and challenge them to speak out. The facilitator ensures that all views and ideas are discussed while simultaneously keeping the desired outcome of the dialogue in mind. In many cases, this means that traditional expert–layperson relations have to be broken down. For example, a facilitator in patient participation must always take into account the traditional role doctor–patient relations or medical expert–layperson relations play.

Interestingly, it is often not only the expert or researcher but also the layperson who maintains the power inequality. Communication research shows that patients are indeed perfectly able to talk meaningfully about research, but they are not always able to express their concerns and ideas effectively (Van der Scheer *et al.*, 2017). That is why it helps to stimulate them to discuss and articulate their views among themselves first before engaging in dialogue with medical experts. In addition, creative techniques and exercises prior to or during the dialogue can create an atmosphere of equality and openness.

4.6 Dialogue in Practice

Organizing dialogue as an open, equal encounter of people is not self-evident. Research on the deliberative activities around the Human Genome Project in the United States shows that the dominant form of science communication still consists of unilaterally informing the public (Brossard & Lewenstein, 2010). In developing countries and in multi-lingual societies, it can be particularly challenging to implement channels and systems for meaningful dialogue between science and publics, as the infrastructure (and possibly the political will) might be lacking. In addition, scientists themselves might resist the inclusion of local or contextualised knowledge in the dialogue as it may be perceived as a challenge to the authority and autonomy of science itself (Kerr, Cunningham-Burley & Tutton, 2007).

In this section, some practical considerations are given that science communicators should take into account when preparing or organizing a dialogue that meets the conditions as described in Section 4.5.5, if they want to arrive at mutual understanding and appreciation of participants as well as reasonable solutions.

4.6.1 *The Goal of Dialogue*

Both scientific and public dialogue participants must be prepared to develop new insights and use them when they return to their daily practice as consumers, urban planners, or researchers. The organizer of a dialogue event has to be clear about the objective of the activity in advance and how that objective is relevant to each of the participants. Is the objective to explore the different perspectives of the participants, or is it intended to reach a joint decision? If the organizer ensures that the matters discussed are relevant to the professional or social community of the various participants, it is more likely that the event will bring real change, both during the interaction and afterwards. If the goal is to influence decision-making, one will have to keep the legitimacy, relevance, and usefulness of the results for policy-makers in mind.

4.6.2 *The Information Flow*

Dialogue is transaction. This means that the participating parties operate both as transmitter and receiver (see also Chapter 3). Therefore, it is important that participants are willing to both listen and speak. For many people, listening is more complicated than it seems. The more people are convinced of their own views and/or the greater their perceived interest is, the harder it is for them to let go of their own ideas and opinions and to open up to other viewpoints.

A facilitator needs to develop an understanding of the possible positions the parties may take and recognize any traditional expert–layperson relationships where the layperson is mainly in a passive role, listening to the expert's explanations. For example, if a dialogue on medical genomics is announced as an 'ask-the-expert-workshop', the traditional doctor–patient relationship will not be averted (Verhoeff & Waarlo, 2013). Instead, patients can be prepared with respect to their active involvement in the dialogue by, for example, being enabled to mutually share and articulate their narratives or viewpoints in advance. A 'professional' representative from a patients' organization who promotes the patients' perspective can also be invited. Issues that may seem to be of minor importance could, on closer examination, be important for increasing mutual involvement. When clinical experts are invited to a location where patients are already at ease, the daily routine of the doctor's office where the doctor explains the diagnosis and treatment and the patient primarily listens or asks questions for clarification will be broken. The reciprocal transfer of information alone does not suffice for a dialogue. It could be a so-called *monologue à deux*. It is also essential that everyone will be able to suggest topics and, thus, help determine the subject of discussion. In addition, all participants should really listen to each other.

4.6.3 *The Dominant Frame of Reference*

In a dialogue, participants exchange views on a specific topic or issue. How this topic or issue is defined depends on their frame of reference. This determines which questions may or may not be asked, how facts are interpreted, and what interests and values are the subject of discussion. In other words the frame of reference defines the boundaries of the topic. This framing of the dialogue is thus essentially a political process (Irwin, 2006; Wynne, 2006). Has the discussion been steered predominantly from one perspective? Has everyone been given the opportunity to tell his or her story? In the Netherlands, social discussions on medical issues have been criticized by the Council for Social Development (Dutch Council for Social Development, 2004) as being dominated by medical experts. As a result, issues were framed in technical medical jargon and less attention was given to social issues in health care and the contribution of patients and patient organizations.

A facilitator of dialogue on science and technology may want to decide what social values and interests should play a role in the discussion. It may be sensible to investigate what the relevant existing perceptions of the problem are and to define the problem in terms of those perceptions. Also, the first step of a dialogue in action may be to define the problem together and create a common ground. To achieve this, it is important for the discussion to start from a shared understanding of the subject and a shared sense of urgency.

4.6.4 *Who is Affected?*

In a social dialogue, all parties with an interest in the issue should be involved. It can be difficult to define who exactly the stakeholders are (Davies *et al.*, 2009). For example, it might not yet be clear how a certain technology will develop and who will be affected by it. The organizer should avoid having the problem and its solution defined by a vocal minority because the majority is silent. With respect to medical issues, it may be considered logical to invite patient organizations representing a directly affected public. They may, however, have specific expectations and desires that focus on new options for diagnosis and treatment without a broader cost–benefit analysis made by (less directly involved) citizens. Organizers can assume a role in which they present the arguments of an absent party. They could, for example, act as the 'devil's advocate' by deliberately advancing counterarguments to explore the full range of arguments in the dialogue. They can also ask other experts, such as ethicists and philosophers, to represent absent or future stakeholders.

4.6.5 *Impact of the Dialogue*

An important criterion for evaluating the success of a dialogue activity is the influence it has afterwards. For non-policy-informing dialogue, the desired impact should be noticeable primarily in the change undergone by the participants. They could display a change in behavior when they return to their daily research, professional practice, or daily life. A change is often hard to trace, but the participants can be asked about their experience right after the dialogue and 2 months later. Have they gained any new insights? And did they apply these in practice? Can they explain how their behavior has changed? Another aspect is that they might feel more equipped to engage in future dialogue. Patients may, for instance, specify that the threshold for speaking with a researcher or physician has been lowered. Conversely, researchers can indicate that they are better equipped to take the patient's quality of life into account when writing research proposals.

A 'learner report' or 'learning progress report' can be a useful tool for participants to reflect on what they have learned or have taken away from a dialogue. Here, the participants complete fill-in-the-blank questions on learning effects to indicate the personal impact of the dialogue event. In addition to these incremental changes, the quality of a dialogue event itself can also be measured by tracing the more visible changes or explicit references to the event in policy documents.

4.6.6 *The Role of the Facilitator*

The previous five points illustrate the importance of a communication professional in preparing as well as guiding the interaction process. Professionals can choose from a wide repertoire of strategic roles (see Table 4.2). They can decide to take a neutral position as a passive chairman and limit their role to ensure that all participants are involved. They can also play a more active or leading role as chairman and pose leading questions and summarize the positions of the participants after a discussion. As was discussed earlier, they can also represent a person or group of people. It may be good to decide picking up this role in advance if a particular party

Table 4.2: The different facilitating roles of the communication professional.

Role of the communication professional	Description
Neutral facilitator/passive chairman	Facilitates the input of the participants and monitors the conditions for dialogue; has no input in the discussion and provides no feedback.
Instructor/active chairman	Provides procedural explanations and content-related information, summarizes the discussions, asks questions, and provides feedback.
Concerned instructor/chairman	States his or her own perspective on the issue and defends it.
Interviewer	Asks questions of the participants to enable them to share ideas, experiences, or feelings.
Devil's advocate	Aims to increase interaction by deliberately presenting and defending contrasting viewpoints.
Observer	Observes the participants and does not take part in discussions.

Source: Adjusted from Harwood (1998).

cannot be present. During the dialogue, the role of 'devil's advocate' can be a way to deliberately insert statements that contrast the viewpoints of any participating party in order to increase interaction. Here, it is important that the communicators are transparent about their intentions so that not too much resistance may be provoked to their role as chairman. An important activity of the communication professionals in stimulating mutual learning may be to clarify or rephrase the viewpoints of participants and make them (more) understandable. One may, for example, act as an interviewer and ask for a more elaborate explanation or reveal the presuppositions that participants may have.

Table 4.2 presents a repertoire from which the experienced communication professional can choose for the purpose of increasing interaction during the communication process. It also provides the possibility for the mediator or facilitator to adapt his or her role to make sure that the dialogue progresses according to the above-mentioned conditions. This role also depends on whether the goal is to reach consensus among the participants or to strengthen mutual learning. In the role of active chairman, for example, a science communicator could act as a bridge between the experts and other stakeholders to reach consensus. Such a role could entail grasping the essential or core ideas of expert knowledge that the other stakeholders need and at the same time understanding the needs, knowledge, and experience of these stakeholders. The science communicator conveys this knowledge to all parties involved so that, together, they can construct new knowledge and find solutions that include all perspectives. Such a model has been used successfully in Mexico to solve environmental problems, including ecological restoration with a sustainable approach (Castillo *et al.*, 2018).

4.7 Conclusion

To find solutions to many social, economic, and environmental problems, the readiness of a scientist to embrace social values and perspectives is more and more seen as essential. The old 'contract' between science and society, in which science communicated its reliable knowledge to society in exchange for autonomy, has gradually being replaced by a reciprocal relation in which science is transparent and participative. Policy-informing dialogue should be distinguished from dialogue to facilitate mutual learning between participants. Although this distinction has important consequences for evaluating the quality or success of dialogue activities, both forms of dialogue foster the credibility, legitimacy, and salience of scientific research and technological innovation.

In practice, it is not self-evident to organize a dialogue that meets important conditions as equality, mutual trust, respect, openness, and understanding. However, meeting these conditions increases the collaborative nature of the dialogue in which participants learn from each other, bring in their expertise, and develop ideas together on a shared issue. In a time where society is increasingly forced to deal with ill-defined, wicked problems and science is contested by 'alternative facts', including relevant lifeworld and professional expertise is key to find solutions that pair reliable knowledge with social awareness and responsibility.

References

Bauer, M. W., Allum, N., & Miller, S. (2007). What can we learn from 25 years of PUS survey research? Liberating and expanding the agenda. *Public Understanding of Science, 16*(1), 79–95.

Brossard, D., & Lewenstein, B. V. (2010). A critical appraisal of models of public understanding of science: Using practice to inform theory. In L. Kahlor & P. Stout (eds.), *Communicating Science: New Agendas in Communication.* New York, NY: Routledge, pp. 11–39.

Burgess, M. M. (2014). From 'trust us' to participatory governance: Deliberative publics and science policy. *Public Understanding of Science, 23*(1), 48–52.

Castillo, A., Vega-Rivera, J. H., Perez-Escobedo, M., Romo-Diaz, G., Lopez-Carapia, G., & Ayala Orozco, B. (2018). Linking social–ecological knowledge with rural communities in Mexico: Lessons and challenges toward sustainability. *Ecosphere, 9*(10), e02470. 10.1002/ecs2.2470

Collingridge, D. (1980). *The Social Control of Technology.* New York, NY: St Martin's Press.

Davies, S., McCallie, E., Simonsson, E., Lehr, & Duensing, J. L. (2009). Discussing dialogue: Perspectives on the value of science dialogue events that do not inform policy. *Public Understanding of Science, 18*(3), 338–353.

Dewey, J. (1927). *The Public and Its Problems.* Chicago, IL: The Swallow Press.

Directorate-General for Research and Innovation (2018). *Europe's Future: Open Innovation, Open Science, Open to the World*. Reflections of the Research, Innovation and Science Policy Experts (RISE) High Level Group.

Dutch Council for Social Development (Raad voor Maatschappelijke Ontwikkeling) (2004). *Recommendation 29: Humane genetica en samenleving. Bouwstenen voor een ander debat.* The Hague: RMO.

Felt, U., Schumann, S., Schwarz, C., & Strassnig, M. (2013). Technology of imagination: A card-based public engagement method for debating emerging technologies. *Qualitative Research, 14*(2), 233–251.

Gadamer, H. G. (1960). *Wahrheit und methode.* Tübingen: Mohr.

Gibbons, M. (1996). *The New Production of Science and Research in Contemporary Societies.* London: Sage.

Gibbons, M. (1999). Science's New Social Contract with Society. *Nature, 402,* C81–C84.

Habermas, J. (1984). *The Theory of Communicative Action.* London: Heinemann.

Harwood, D. (1998). The teacher's role in democratic pedagogies. In C. Holden & N. Clough (eds.), *Children as Citizens, Education for Participation.* London: Jessica Kingsley Publishers, pp. 154–170.

Hisschemöller, M., & Hoppe, R. (1996). Coping with intractable controversies: The case of problem structuring in policy design and analysis. *Knowledge and Policy: The International Journal of Knowledge Transfer, 8,* 40–60.

Hisschemöller, M., Hoppe, R., Dunn, W., & Ravetz, J., (eds.), (2001). *Knowledge, Power, and Participation in Environmental Policy Analysis.* New Brunswick, NJ: Transaction Publishing.

Irwin, A. (2006). The politics of talk: Coming to terms with the 'new' scientific governance. *Social Studies of Science, 36,* 299–320.

Kerr, A., Cunningham-Burley, S., & Tutton, R. (2007). Shifting subject positions: Experts and lay people in public dialogue. *Social Studies of Science, 33*(7), 385–411.

Rittel, H. W. J., & Webber M. M. (1973). Dilemmas in a general theory of planning. *Policy Sciences, 4,* 155–169.

Smaling, A. (2008). *Dialoog en empathie in de methodologie.* Utrecht: Humanistics University Press.

Van der Scheer, L., Garcia, E., van der Laan, A. L., van der Burg, S., & Boenink, M. (2017). The benefits of patient involvement for translational research. *Health Care Analysis, 25*(3), 225–241.

Verhoeff, R. P., & Waarlo, A. J. (2013). Good intentions, stubborn practice: A critical appraisal of a public event on cancer genomics. *International Journal of Science Education Part B, 3*(1), 1–24.

Von Schomberg, R. (2007). From the ethics of technology to the ethics of knowledge assessment. In P. Goujon, S. Lavelle, P. Duquenoy, K. Kimppa, & V. Laurent (eds.), *The Information Society: Innovation, Legitimacy, Ethics and Democracy*, Vol. 233, In Honor of Professor Jacques Berleur S. J. Boston, MA: Springer, pp. 39–55.

Wilsdon, J., & Willis, R. (2004). *See-through Science: Why Public Engagement Needs to Move Upstream.* London: Demos.

Wynne, B. (2006). Public engagement as a means of restoring public trust in science — hitting the notes, but missing the music? *Public Health Genomics, 9*(3), 211–220.

Chapter 5

Informal Science Education

Anne M. Land-Zandstra, Liesbeth de Bakker,
and Eric A. Jensen

5.1 Introduction

Learning about science is a lifelong process and can happen anywhere. When it is organized by schools, it is called formal science education. When it happens outside the school system, it is called informal science education. A major part of science learning is not tied to schools (National Research Council, 2009) but takes place in organizations such as science museums, science festivals, zoos, and nature parks.[1] Such institutions develop special services to promote informal science learning.

The important aims of informal science education include sparking an interest in science, enhancing science-related knowledge, reasoning and skills, and promoting pro-science identities. A pro-science identity encompasses a positive orientation toward evidence-based approaches to knowledge and a feeling of empowerment to participate in discussions about scientific issues (National Research Council, 2009).

Societal trends are driving the need for and an interest in informal science education. These trends include the increasing complexity of the economy and societal structures, with scientific and technical issues often closely involved. In this context, people need new skills and knowledge to thrive. Some of these new skills and knowledge domains are under-developed

---

[1] Informal science education can also be delivered through mediated communication platforms such as radio, television, and the Internet (e.g. including Massive Open Online Courses). However, this chapter focuses on the face-to-face formats for informal science education.

in formal education systems, such as handling large information streams, critical thinking, working in teams, and, to a lesser extent, constructing knowledge and creativity (Binkley *et al.*, 2012). In addition, Europe faces declining numbers of students interested in pursuing science-related careers (Sjøberg & Schreiner, 2010). On both counts, professionals and policy-makers turn to informal science education for solutions.

Several reasons for the importance of informal science learning can be identified. First, engagement with this process by adults is often driven by personal and professional needs for information and technical skills that emerge from daily life. Second, informal science learning may also be motivated by leisure interests, that is, some people enjoy learning more about scientific phenomena in a variety of settings such as science museums, science centers, science festivals, or citizen science projects. Third, informal learning institutions sometimes offer experiences that schools cannot offer (e.g. seeing live animals in a zoo or meeting practicing scientists). Particularly, in some countries where the formal school system does not offer a lot of science education, informal science education activities may serve as additional ways for young people to learn about science. Fourth, some informal science education programs are motivated by the hope that young people can get interested in science as a subject area and a career opportunity through these experiences. Informal science institutions can help spark and increase excitement for science and technology, sometimes overcoming negative experiences with school science identified as boring and irrelevant (Stocklmayer, Rennie & Gilbert, 2010). Finally, a less common motivation (Jensen & Holliman, 2016) for informal science learning practice is the fact that some scientific knowledge is needed to be able to take part in democratic debates about scientific and technological issues. What developments should be allowed? Which ones shouldn't? What is desirable?

So, in order to get a better understanding of this subdomain of science communication called informal science education, this chapter explains what informal science education is and how it differs from formal science education. It identifies the core strengths and characteristics, theoretical perspectives, and current debates relevant to this field. In addition, it gives examples of institutions and programs for which informal science education is their core business.

5.2 Formal and Informal Science Education

Informal science education and formal science education are both part of the broader domain of science education. Although a sharp distinction may be artificial, as will be discussed in what follows, it helps to first consider the differences between the two. Formal science education refers to learning and teaching about science within the school system, bounded by standards, curriculum, lesson plans, and tests. Learning about science outside of this system is called informal science education. Examples of spontaneous informal science education include looking up information online about the effectiveness of a new type of diet, sitting at home watching a nature documentary on television, or reading an article about a medical breakthrough in a newspaper or science magazine. One can also experience informal science education in designed settings such as museums, science festivals, botanical gardens, zoos, aquariums, and nature centers or in science cafés, citizen science projects, or after-school clubs.

Important defining characteristics of informal science education include the following:

- is often based on voluntary participation;
- provides opportunities for social interaction with other participants;
- connects to personal interests through a learner-centered approach;
- lacks formal assessment.

In contrast, typical characteristics of formal science education include the following:

- compulsory participation;
- formal structure;
- working according to a curriculum;
- formally assessed with tests and grades.

Table 5.1 contains a somewhat older, but broadly accepted, list of characteristics which can be used to discern formal from informal science education (Hofstein & Rosenfeld, 1996; Wellington, 1990).

Of course, this kind of sharp distinction between formal and informal science education is less clear in practice. Both formal and informal science

Table 5.1: Differences between formal and informal science education.

Formal science education	Informal science education
Compulsory	Voluntary
Structured, sequenced	Unstructured, unsequenced
Assessed	Non-assessed
Close-ended	Open-ended
Teacher-led	Learner-led
Classroom and institution based	Outside of formal settings
Curriculum based	Non-curriculum based
Fewer unintended outcomes	Many unintended learning outcomes
Social aspect less central	Social aspect central
Empirically measured outcomes	Less directly measurable outcomes

Sources: Based on Wellington (1990) and Hofstein & Rosenfeld (1996).

education are typically designed by professionals (teachers and informal science educators), who make decisions based on a wide range of influences (including theoretical, as noted later in this chapter). In both cases, there are varying levels of structure and scope for learners to make their own decisions about what they would like to do and learn.

Hybrid learning experiences blend aspects from both formal and informal science education. For example, teachers may organize workshops in their classrooms as part of the formal curriculum with learner-directed and so-called *free-choice learning* to achieve their learning aims. Another example that resides somewhere between formal and informal is a field trip to a science museum. When this visit is strictly planned and mostly takes place within a separate room following a tight structure, it looks a lot like formal science education. In practice, learning in different contexts is cumulative. Every learning opportunity, inside a school or at the kitchen table, can contribute to science learning.

In formal as well as in informal science education, learning is the core aim. In this chapter, learning is defined not just as the acquisition of factual knowledge about science but as encompassing a broader range of possible outcomes including increased interest and change of attitudes. In formal science education, the main focus is on acquiring knowledge and skills.

In informal science education, often more attention is given to increasing interest in science, developing positive science attitudes, and changing behavior. Here, learning outcomes are more oriented toward personal development and growth. This is where informal science education may be particularly suitable to complement formal science education. However, measuring the impact on interest, attitude, and behavior is a lot harder than measuring changes in knowledge and skills because these variables are more stable and many other factors besides the visit or interaction may have an impact.

For the purpose of describing and measuring this broader range of learning outcomes, several organizations have developed frameworks for learning in (informal) science education (Friedman, 2008; Hooper-Greenhill *et al.*, 2003; National Research Council, 2009). For example, according to a National Science Foundation (NSF) report, science learning should include knowledge, interest in science, attitude toward science, behavior, and skills (Friedman, 2008).

According to this NSF report, the dimension of knowledge includes the understanding of topics, concepts, phenomena, theories, and careers of science. The dimension of interest is of importance because when people are interested in science, they may want to learn more about it, or they may have a more positive attitude toward scientific developments. Interest in science can even lead to the pursuit of a scientific career (Hidi & Renninger, 2006). Attitudes toward science include a combination of relatively stable feelings, beliefs, and values related to science (Ajzen, 2001; Van Aalderen-Smeets, Walma van der Molen & Asma, 2011). Science learning can also include a change of behavior. For example, a learning experience can cause someone to start recycling or to start eating healthier, although the link between learning and practical behavior changes may be very weak (Moss, Jensen & Gusset, 2017). Lastly, science learning can include acquiring science-related skills, for example, developing hypotheses, working with research instruments such as microscopes, or computer software and data analysis (Friedman, 2008).

A widely praised method of developing these types of learning in both formal and informal settings is *inquiry-based learning*. This can be defined as a process of discovering new causal relations, with the learner formulating hypotheses and testing them by conducting experiments and/or making observations (Pedaste *et al.*, 2012). Participants can acquire scientific knowledge and skills through observing, hypothesizing, collecting data, discussing findings, and drawing conclusions. Understanding this idealized model of the scientific process is an important part of science learning. Inquiry-based learning can be effective because people generally learn more easily and enthusiastically when the learning activity springs from their own interest and when it answers their own questions (Hohenstein & Moussouri, 2018). Box 5.1 describes an example of what inquiry-based learning in a science museum can look like.

Box 5.1:   Was Archaeopteryx able to fly?

Photo credit: Leonora Simony.

The Natural History Museum of Denmark in Copenhagen offers a paleontology-related school program for upper secondary school students (16–18 years old) called Evolution: From Dinosaur to Bird. The inquiry-based learning segment of this program includes a 40-minute exercise during which students compare a modern bird skeleton to a cast of a fossil Archaeopteryx. Just like in real evolutionary science, students are motivated to make observations of similarities and differences that form the basis of answering questions about evolution and the possible link between dinosaurs and birds.

Students worked in groups of four with, for instance, a bird skeleton of a crow, a pigeon, or a gull and a cast of an Archaeopteryx fossil. The students

---

Box 5.1: *(Continued)*

were guided by a single overarching research question: was Archaeopteryx able to fly? Other than that, the groups could decide how to address that question by themselves. They carefully observed the similarities and differences between the modern bird skeleton and the fossil cast. The research question triggered the students to follow the different steps of authentic inquiry. As a result, they formulated hypotheses, studied aspects like feather size, wing area, and presence of claws; made comparisons, discussed findings, and drew conclusions.

A study focusing on the exercise itself showed that in more than half of the groups studied, the students were able to work and think like real paleontologists. The skeletons and the fossil cast, and the presence of a specific research question 'Was Archaeopteryx able to fly?' clearly helped the students with that (Achiam, Simony & Lindow, 2016). This means that the students indeed had worked like scientists and had learned science-related skills through this particular exercise based on inquiry-based learning. In this example, students had no free choice in picking the program, the objects, or the research question, but they were free to decide how to conduct their research.

Another approach to develop science-related skills which has grown tremendously in popularity over the past decade is *design-based learning* (e.g. in the form of tinkering or maker education). This approach has especially been welcomed by educators in informal settings such as libraries, after school science clubs and museums. In tinkering or maker spaces, participants give themselves a challenge. They want to build a solution to a problem they have defined themselves using the materials and tools available in that particular maker space, such as cardboard, sensors, batteries, hot-glue guns, three-dimensional printers (Gutwill, Hido & Sindorf, 2015). The activity is therefore often learner-driven and allows participants to learn by doing (Vossoughi & Bevan, 2014).

To get an idea about the learning behaviors that take place in a tinkering or maker space, the San Francisco Exploratorium studied and made videos of participants in its on-floor Tinkering Studio (Gutwill, Hido & Sindorf, 2015). Analysis of the video materials showed that four Dimensions of

Learning could be discerned: engagement (with the task at hand); initiative and intentionality (setting goals and trying to meet them); social scaffolding (willingness to help others present); and development of understanding (knowing what is happening and working toward achieving the aim). These behaviors indicate the potential of Maker education as a form of science learning.

5.3 The Power of Informal Education

The unique characteristics of informal science education can help young and old to gain new perspectives on science. A commonly identified characteristic of informal science education is the opportunity for so-called *free-choice learning* (Falk & Dierking, 2002). While many choices are constrained in informal science education, either by the institution or the participant's background, there may be more room for deciding where to visit, which activities to join, the items to view, and the experiences to undergo (Hohenstein & Moussouri, 2018). This relatively flexible structure may facilitate visitors' search for personal relevance within the learning experience. Because visitors can make more choices, they can seek out experiences that connect to their lives and interests. Personal relevance and interest influence attention, learning goals, and levels of learning (Hidi & Renninger, 2006). People who are more interested in a particular topic have better learning outcomes, show higher perseverance when difficulties arise, and develop deeper levels of learning than less interested people.

Another characteristic of visits to informal learning settings such as museums and zoos, for example, is that there is room for unstructured social interaction because many people visit these institutions within a group. During their visit, people learn while talking and collaborating with their group members and facilitators from the institution (Moss, Jensen & Gusset, 2017), which has a positive effect on learning (Falk & Dierking, 2013).

Another potential strength of informal education is that visitors may encounter real objects, real phenomena, real scientists, and real science (Braund & Reiss, 2006). Many people visit informal science learning institutions in order to see the real thing (Pekarik, Doering & Karns, 1999). Many museum educators are convinced that providing these authentic experiences is important. Some evidence exists that real objects can spark interest and amazement that can lead into learning (Bunce, 2016; Van Gerven, Land-Zandstra & Damsma, 2018). However, comparative studies that systematically compare the impact of real objects vs models are scarce. In addition, interacting with real scientists and engaging in real scientific research, for example, through certain kinds of science festivals, science cafes, open days of research institutes, or citizen science initiatives, may improve visitors' or participants' knowledge about and opinion on science (Jensen & Buckley, 2014; Land-Zandstra *et al.*, 2016).

5.4 Theoretical Views to Guide Informal Learning Practice

Although the field of research about informal science education is growing, in reality, theory rarely features in the practice of informal science education. A clear need for informal science educators to better integrate theory and practice is felt, as was also addressed in the National Academy of Sciences report on *Learning science in informal environments* (National Research Council, 2009):

> There is an interdisciplinary community of scholars and educators who share an interest in developing coherent theory and practice of learning science in informal environments. However, more widely shared language, values, assumptions, learning theories, and standards of evidence are needed to build a more cohesive and instructive body of knowledge and practice. (p. 305)

While theoretical perspectives from fields such as sociology and psychology have started to gain greater recognition within research on informal science education, the models emerging from within the field have tended have tended to simply offer basic categories for understanding informal learning. For example, the *contextual model of learning* by Falk & Dierking (2000) identifies three contexts that form and influence learning experiences in informal settings: the personal, sociocultural, and physical context. Figure 5.1 shows an overview of the three contexts of learning.

Lynn Dierking and John Falk distinguish three contexts that together form the museum experience and influence the learning process.

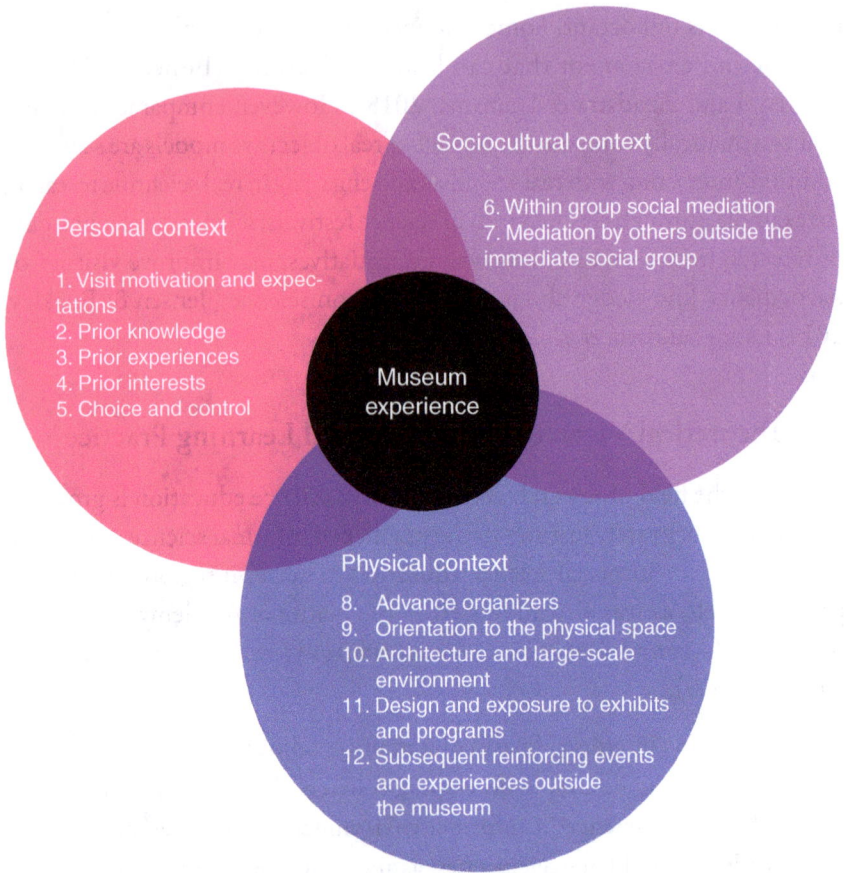

Personal context

1. Visit motivation and expectations
2. Prior knowledge
3. Prior experiences
4. Prior interests
5. Choice and control

Sociocultural context

6. Within group social mediation
7. Mediation by others outside the immediate social group

Museum experience

Physical context

8. Advance organizers
9. Orientation to the physical space
10. Architecture and large-scale environment
11. Design and exposure to exhibits and programs
12. Subsequent reinforcing events and experiences outside the museum

Figure 5.1: Falk and Dierking's Contextual model of learning.

First, the personal context of the individual visitor is important. What do visitors want, know, and feel regarding their visits, and how much freedom do they have in designing their own experiences? Within this context, interest, motivation, and expectations are important elements. Second, the sociocultural context includes with whom the visitor comes, what is being discussed, in what way, what kind of interaction is taking place among friends and family, as well as with other visitors and guides. The sociocultural context also considers the visitors' social background which influences how and what they learn. Third, the physical context applies to the physical space where the experience takes place, the building, the activities, and even events that happen before or after the visit that are related to the informal learning experience (Falk & Dierking, 2000).

According to Falk & Dierking (2000), within each context, developers of informal learning experiences can make several choices to develop successful experiences for their visitors. Within the personal context, informal experiences should connect to personal experiences and visitor expectations. Within this context, knowledge about learning processes is important. Another important aspect is connection of the experience to participants' interests and increasing enthusiasm. In terms of the sociocultural context, developers can ensure that during learning activities space is available for social interaction and learning from and with others. Museums often stimulate social interaction by building exhibits that entice visitors to collaborate. Another important aspect in this context is making experiences relevant for people with diverse cultural backgrounds. Finally, within the physical context, it is important that the physical space stimulates learning, or at least does not inhibit it. For example, visitors should feel comfortable, know where to find restrooms, find a *hands-on exhibit* easy to operate, or know how to navigate the museum. The physical context also builds on experiences that people have outside of the institution.

While the points made in the *contextual model of learning* are long-established within other disciplines such as communication and psychology (e.g. Spitzberg, 1983; Maslow, 1943), many informal science educators find it helpful to have these categories spelled out in terms of the informal learning context. Theoretical perspectives from these other disciplines often provide more depth and explanatory power regarding the aspects that are important within each context (e.g. Vygotsky's learning theory about social interaction, as explained below). Although for Falk, the visits remain the site of primary interest, other researchers contend that the primary focus for informal science education should be on visitors' lives in which they connect different learning experiences. Informal science educators should therefore not forget they can also draw on a wealth of theory and research from sociology, psychology, and communication (Dawson & Jensen, 2011; Lemke, 2000). The field of informal science learning lacks a single overarching, integrative theory, but depending on the goals and characteristics of an informal learning experience, different theories may be useful for informing practice.

5.4.1 *Relevant Learning Theories*

From the beginning of the 20th century, several theories of learning have been developed that still influence present views of learning. These theories can roughly be divided into three theoretical perspectives: behaviorism, cognitivism, and sociocultural theory (National Research Council, 2009). Proponents of behaviorism define learning mainly as behavior change. They focus on what the learner can do at the end of a learning activity. In this view, learning is a process of rewarding correct behavior and discouraging all other behavior. Cognitive theories (based on the work of Piaget, 1936; Ausubel, 1968) focus more on how people construct knowledge structures. The influence of prior knowledge plays an important role. When learners acquire new information, they try to connect it to what they already know or can do. Sociocultural theories (tracing back to the work of Bandura, 1977; Vygotsky, 1978) also highlight how people build on existing knowledge, while emphasizing the social aspect of learning. That is, the social surroundings of the learner are important. Learning is affected by the people a person learns with and from, and past interactions. Learning through explicit guidance, observation, and imitation are pathways within this perspective.

For informal science education practitioners, cognitive and sociocultural theories are most relevant, including important concepts such as experiences and (inter)actions (National Research Council, 2009). Within an informal learning environment, visitors' active engagement can be supported through well-designed activities. Furthermore, prior knowledge of visitors has an important impact on the informal learning experience. What people know about a certain topic influences the way they acquire new information while visiting a museum or zoo. In addition, learning by social interaction plays an important role in informal science education, as can be demonstrated by the application of the work of Vygotsky to an informal learning experience.

In terms of relevant sociocultural learning theory, influential psychologist Vygotsky (1978, 1994) showed that human thinking, memory, imagination, and attention develop out of social interaction, rather than being purely individual (see also Vygotsky & Luria, 1994). Vygotsky's theory has important implications for informal learning, by emphasizing that the facilitator (whether that is a person, object, information panel, video, etc.) plays a vital role in drawing a visitor's attention in useful directions that support learning. Following this theory, informal learning institutions should focus on providing conceptual tools that allow visitors to develop their knowledge beyond what could be achieved autonomously. In other words, this theoretical model of learning places informal science educators in the role of toolmakers, fashioning the most effective concepts and explanations possible and then provisioning visitors with these concepts for them to use to leverage themselves into a higher level of learning. One metaphor that is often used for this process is *scaffolding*, to highlight the need to provide structural, social, and conceptual support for learning. Box 5.2 shows an example of how scaffolding can work in practice.

Box 5.2: Evidence of kids' learning at the zoo

Support for Vygotsky's learning model and the effective use of scaffolding in informal science learning has been found in an annotated drawing study in the context of London Zoo (Jensen, 2014). With effective social learning provision, either by their own teachers or by zoo staff members, students (aged 7–14) were capable of observing for themselves that there was a diversity of types of animals, which may be greater than they had previously thought. This type of learning even when provided with only limited scaffolding is seen when children's interests go beyond charismatic mammal species such as lions and polar bears to include a broader diversity of animals (see change evident in the matched drawings for an 8-year-old child in Figure 5.2).

In the same study, there was evidence that informal science educators on staff at the zoo were better able to scaffold the learning of schoolchildren than their own teachers. They did this by providing conceptual resources, or specific ideas, relevant to the zoo context in order to help children see how the animals in the zoo related to wild animals in distant habitats. This addition of new ideas into a child's existing understanding of animals and their habitats was enhanced through facilitation, in line with Vygotsky's (1978) theory. In other words, with greater communication about relevant scientific concepts linked to the zoo experience, informal science educators at the zoo were able to extend the impact of this informal science learning experience.

(Continued)

Box 5.2: (*Continued*)

| BEFORE | AFTER |

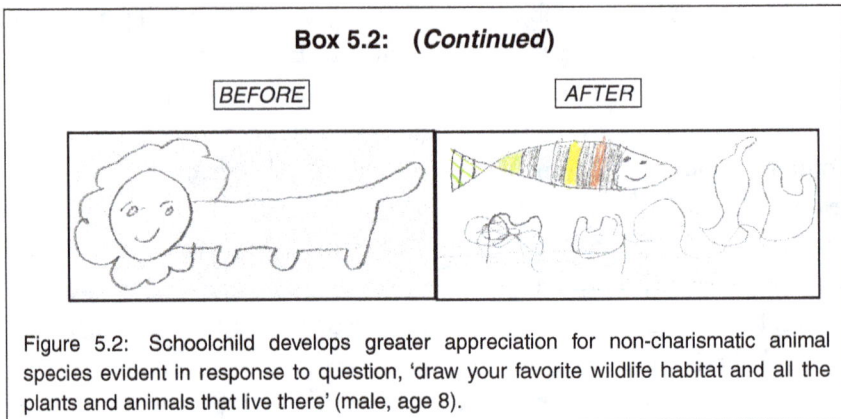

Figure 5.2: Schoolchild develops greater appreciation for non-charismatic animal species evident in response to question, 'draw your favorite wildlife habitat and all the plants and animals that live there' (male, age 8).

5.4.2 *Identity-related Motivations*

In addition to looking at the learning process within informal environments, it is relevant to consider why visitors or participants engage in learning experiences in the first place. Their motivations may relate to their behavior and learning (Packer & Ballantyne, 2002; Moss, Jensen & Gusset, 2017) and can be considered when designing informal learning experiences. Visitor groups for informal science experiences are often not homogeneous; many different types of people visit these settings. Therefore, it is important to provide relevant experiences for different visitor groups. Falk (2009) identifies five categories of *identity-related motivations*, claiming these motivations depend on visitors' identity at the moment of the visit. Falk argues that these identities are not fixed but depend on the specific situation. The five categories are explorer, facilitator, experience seeker, professional or hobbyist, and recharger.

According to Falk (2009), *explorers* are driven by curiosity; they like to discover new things and learn more about a variety of topics and are mostly concerned about their own experience and not so much about their companions. Facilitators, in contrast, are driven by a desire to fulfill the needs of the people they are with, often their (grand)children. They want their children to have a good time and a fulfilling experience. For them, the focus is making the most out of their children's visit. The *professional/hobbyist* category comprises visitors who have a high level of knowledge about a certain topic, which they develop further through the museum visit. That is, the visit for this category of visitor relates to her or his hobby or profession, meaning depth of content will be needed to satisfy this type of visitor. *Experience seekers* are defined by Falk as being motivated to have 'been there, done that' by visiting the museum. Finally, the *recharger* wants to fulfill a more spiritual desire for contemplation or reflection, according to Falk's categorization. Rechargers need quiet places to look, think, and reflect about an object or installation.

Falk downplays the importance of what he calls 'big "I" identities', such as age, geography, ethnicity, and social class. Instead, he argues it is what motivational category the visitor fits into that matters. Moreover, he acknowledges some potential for change. A family group can consist of many experience seekers, going their own way, or a facilitating parent who wants to provide a good experience for his children. One person can also have a certain type of motivation one day and come back to the same museum or event the next day with a completely different motivation. Falk (2009) argues that the types of motivations are associated with the behavior of people during their visit and the long-term learning impact of the visit.

However, Falk's identity-related motivations (2009) have been criticized for a number of reasons (Dawson & Jensen, 2011). Dawson and Jensen critique audience segmentation approaches in general, and in particular Falk's identity-related motivations model of museum visiting as problematic. They argue that research should place museum visits within a holistic and long-term framework of individual life circumstances, relationships, and trajectories, rather than placement into one of these five segments, because those do not take into account the complexity of informal learning experiences. In any case, whether Falk's visitor categories are accepted or not, it is clear that taking the motivations of visitors into consideration can improve the design and effectiveness of informal science experiences.

5.4.3 *Social Inclusion and Equity*

Another important issue within the informal science education field is the issue of *social inclusion and equity*. Who has access to informal learning opportunities and, more importantly, who does not? Indeed, a long-standing problem of social inequality in informal science education exists, wherein those from more privileged backgrounds are more likely to participate in these types of experiences (Kennedy, Jensen & Verbeke, 2017). Moreover, evidence that informal science educators have unintentionally designed the content and structure of their experiences in a way that advantages those who come from economically and educationally advantaged backgrounds is found (Dawson, 2014).

Overwhelmingly, informal science education organizations and programs are serving a white, middle-class or upper class, highly educated audience (Dawson, 2014). But, even if informal science institutions were able to attract more diverse visitors, their programs and exhibitions may not be satisfying to these newer audiences because of culturally specific ideas of what science learning is. Non-visiting groups may not feel like these opportunities are for them (Feinstein, 2017).

Dawson (2014) proposed a framework for inclusion and equity in informal science education, consisting of three dimensions. First, the issue of *infrastructure access* needs to be considered, both physically in terms of location and entrance fee and less literally in terms of marketing, staff hiring, and involvement in program development. The second dimension is *literacy*, or understanding the rules of the game of science and of the institution itself (e.g. knowing how interactive science exhibits usually work). The third dimension is *community acceptance*: the community of practice needs to start accepting the underrepresented audiences and make them feel welcome. Informal science institutions and programs should work on all three levels of inclusion and equity to truly become available for all publics.

A theory that can help understand the role of museums and other cultural institutions (including informal science education institutions) in social inequality is French sociologist Pierre Bourdieu's *Theory of Practice*. This theory helps to explain how inequality persists despite the efforts of governments and institutions to create a more equal society (Bourdieu, 1986). Bourdieu's concept of *cultural capital* refers to a type of non-economic resource, a combination of knowledge, habits, interests, and behaviors associated with economic and cultural elites. This resource is passed down primarily through families, for example, from parents to their children. This resource is used to distinguish between different categories of people, excluding the less educated or less cultured individuals from participation. This theory and the concept of cultural capital have valuable applications to informal science education for helping to understand how to ensure it is not exclusionary.

A more recent contribution by Archer *et al.* (2015) is to introduce the concept of *science capital*, which like cultural capital can be used to separate 'us' from 'them' in society. However, it is important to bear in mind that there is a 'systemic pattern throughout society reproducing deeply unequal, unjust and exclusionary social relations, even when financial barriers to cultural participation appear to have been addressed (for example, through free entry to museums and galleries)' (Jensen & Wright, 2015, p. 1144). This means that informal science educators must work particularly hard to avoid having their institutions become just another part of this larger unequal system. For example, if informal science education is more easily available and more commonly accessed by already privileged individuals, or if they are designed in a way that only those with pro-science parents will benefit, then it could be contributing to social inequality rather than reducing it. As discussed earlier, inclusion of underrepresented visitors has received increased attention recently. According to Simon (2016), the key is for informal science learning institutions to establish their relevance to the community and create meaningful experiences that can be understood and accessed by all.

5.5 Examples of Informal Science Education

Basically, any experience or activity through which people can learn about science outside of the formal school system is considered informal science education. These experiences can be roughly divided into three categories: everyday settings, informal science learning institutions, and organized programs for informal science education (National Research Council, 2009).

Informal science education in everyday settings can take different forms and often does not happen with the specific goal of learning about science. Sources of science education in everyday settings include radio and television programs, science-related websites, science blogs, books or magazines, social media, and applications for smartphones or tablets. In particular, technological developments in the areas of Internet and social media make it possible for people to have scientific knowledge available at their fingertips wherever and whenever they want.[2]

Informal science learning institutions include science museums, zoos, science centers, libraries, national parks, and nature centers. These institutions are often founded with informal science education as a key goal. Visitors often have a relatively free choice about what to do within these institutions. Within the category of informal science learning institutions, science museums and centers play an important role. During the mid-1970s a large shift occurs in the field of science museums from the traditional role of protectors of heritage toward institutions that shift from an object focus to more interactive exhibition designs. Around the same time, many museums expand their dedicated education departments. In an attempt to effectively meet the needs of the increasingly diversifying audiences and pressures of modern-day society, museums invest in approaches that aim to further engage the public with science. For example, through hands-on science exhibits, science centers increasingly focus on inquiry-based learning, or maker education, in an attempt to increase involvement of visitors.

The third group, organized programs for informal science education, has the explicit goal to teach people something about science or to get them interested in science, just like informal science institutions. Examples of such programs include science clubs, science festivals, open days of scientific institutions, workshops, lecture series, science cafes, and science debates. An example of organized programs for informal science education that has seen a rapid growth over the past 10 years and can enjoy an increasing popularity and attention is called *citizen science*.

5.5.1 *Citizen Science*

Citizen science is a label that applies to projects that have educational outcomes at their heart and where citizens are actively involved with different

---

[2] As this chapter focuses on face-to-face formats for informal science education, this category of informal science education will not be further elaborated on.

aspects of the scientific process on a voluntary basis. For instance, they can take measurements, collect data, or analyze data (Bonney *et al.*, 2009). Examples of citizen science are projects in which citizens are asked to observe and count different animals or plants, such as bird counts, butterfly counts, or to report the first bloom of specific plants. Other examples are online projects in which citizens analyze data that are too extensive, too time-consuming, or too complex for scientists' computers, such as classifying galaxies, analyzing light curves of stars, or detecting exoplanets. Besides the prevalent data collection and data analysis projects, some projects exist in which citizens are involved in the planning and development stages of scientific studies. In citizen science projects, scientific goals of data collection and analysis are combined with informal science education goals of getting in touch with, learning about, and being actively involved in science.

Although citizen science is not new, it has grown as a field in the last few decades, partly because of the growing opportunities for communication and sharing of data that technologies such as smartphones and the Internet provide. Some overarching principles have emerged to define what citizen science entails (European Citizen Science Association, 2015). These principles include that citizen science should engage citizens in genuine scientific projects that generate new knowledge and understanding; both the scientists and the citizens should benefit in some way from the project; citizens receive feedback about their contributions and those contributions are acknowledged in results and publications.

Bonney *et al.* (2009) identify three different types of citizen science projects. In *contributory* projects, the majority of citizen science projects, scientists decide on the protocols, and participants contribute data or analyses. Examples of such projects are bird counts, water quality projects, or online analysis projects, such as identifying animals through camera trap pictures. An example of a contributory project is the Galaxy Zoo project where participants identify online pictures of galaxies.

In *collaborative* projects, participants are involved in other steps of the scientific process as well. They may help interpret the data, draw conclusions, or disseminate the data. An example of a collaborative project is the Cyanobacteria Monitoring Collaborative, a water quality project in the USA. Participants collect data about the bloom of cyanobacteria, help interpret the data, and help predict when and where outbreaks of the bacteria will occur. The information from the project can be used on local and national scales to inform the measures that are taken to increase water quality.

Finally, in *co-creative* citizen science projects, participants act alongside scientists and are involved in all phases of the scientific process, from coming up with the research questions and developing the research method, to collecting, analyzing, and interpreting the data. Box 5.3 describes an example of a co-creative citizen science project. Such co-creative projects

---

Box 5.3:   Citizens check air quality

Text: Yaela Golumbic, science communication researcher, Technion — Israel Institute of Technology.

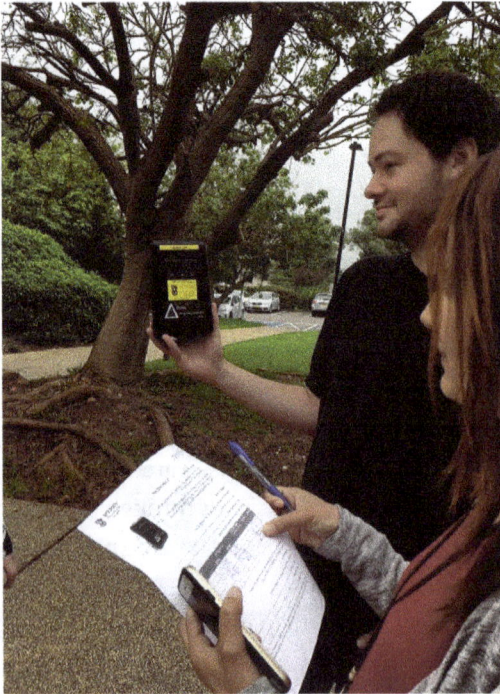

Photo credit: Yaela Golumbic. Two participants of 'Sensing the air' conducting air quality measurements and recording sensor location.

Box 5.3: (*Continued*)

Sensing the Air (https://air.net.technion.ac.il) is an example of a citizen science project with a strong co-creation emphasis. It originated as one of the seven case studies of the European FP7 project CITI-SENSE for developing sensor-based citizen observatories. Sensing the Air aims to facilitate air quality research through active involvement of volunteers and through the collection and interpretation of meaningful air quality data. It further facilitates the access of air quality data to non-experts, as it presents real-time air quality information (obtained from project sensors and official governmental monitoring), in a clear and simple online platform.

The project demonstrates open-ended participation in citizen science as it provides participants small mobile sensors for conducting personal air quality research. Participants use these devices to measure air quality in their local environment and in places of interest (such as their homes, schools, and offices), and explore relevant and personal day-to-day questions regarding air quality. Examples for such questions include: Which is the cleanest room in my home? Should I open the window in the baby's room today? What way should I walk my kids to school?

Participants in the project exhibited a number of motivations for participating in Sensing the Air. These included worried residents who wanted to understand the air quality status in their close environment, activists who were looking for information to support their civil actions, and educators who wanted to engage their students in authentic inquiry.

A study of learning outcomes among participating students (Golumbic, Baram-Tsabari & Fishbain, 2016) revealed an elevation in student content knowledge regarding air quality and an increased understanding of scientific practices. Participation has also shown to increase awareness towards the environment and specifically to topics of air quality and air pollution. Yet, these gains were limited to the topics discussed with the student during participation and did not extend to deeper understanding of the nature of science. Prolonged work with students, emphasizing the nature of science and setting goals for higher level thinking, could further promote critical thinking and a more general understanding of what science is.

are scarce but allow for a true collaboration between citizens and scientists with each party learning from each other and taking into account local knowledge. The research questions in such projects often arise from people's everyday situation, for instance, is it safe to grow vegetables in my community garden that is close to contaminated grounds? Scientists collaborate to provide citizens with rigorous scientific methods of collecting and analyzing data.

Citizen science projects not only contribute to science and research but also provide unique possibilities for informal science education (Bonney *et al.*, 2009). Participants usually join these projects out of personal interest. They want to learn about scientific topics. At the same time, sometimes without being aware of it, participants acquire knowledge and skills in the area of the scientific process because they are personally involved in that process. In addition, citizen science projects can contribute to developing an interest and engagement in science.

5.6 Conclusion

Informal science education is a varied field of practice, with a continuously developing knowledge base (Patrick, 2017). Informal science education differs from formal science education in that it takes place outside of school structures. The definition of informal science learning is broad, although acquiring knowledge and skills, creating interest, improving attitudes toward science and changing behavior are widely shared goals.

In order to realize the potential of informal learning, educators need to account for a number of points. First, there should be a balance between so-called free-choice and guidance for participants (McComas, 2006). Second, informal science educators need to be equipped with the knowledge and skills to effectively support visitors through these experiences. Informal science education requires listening to visitors in order to facilitate the development of their learning around their personal interests and curiosity. In particular, in the case of inquiry activities, guides need to develop their facilitating skills to effectively support learning (Petrich, Wilkinson & Bevan, 2013). Moreover, major limitations exist in terms of the social inclusion and accessibility of informal science education spaces that need to be addressed.

Informal science education's potential to contribute to lifelong learning is still gaining recognition. As informal science educators gain skills and insights about relevant theory and ongoing research, they can sharpen their practice and enhance their impacts.

References

Achiam, M., Simony, L., & Lindow, K. B. E. (2016). Objects prompt authentic scientific activities among learners in a museum programme. *International Journal of Science Education*, 38(6), 1012–1035. doi: 10.1080/09500693. 2016.1178869.

Ajzen, I. (2001). Nature and operation of attitudes. *Annual Review of Psychology*, 52, 27–58.

Archer, L., Dawson, E., DeWitt, J., Seakins, A., & Wong, B. (2015). "Science capital": A conceptual, methodological, and empirical argument for extending Bourdieusian notions of capital beyond the arts. *Journal of Research in Science Teaching*, 52(7), 922–948. doi: ORG/10.1002/TEA.21227.

Ausubel, D. P. (1968). *Educational Psychology: A Cognitive View*. New York, NY: Holt, Rinehart & Winston.

Bandura, A. (1977). *Social Learning Theory*. Englewood Cliffs, NJ: Prentice-Hall.

Binkley M., Erstad, O., Herman, J., Raizen, S., Ripley, M., Miller-Ricci, M., & Rumble, M. (2012). Defining twenty-first century skills. In P. Griffin, B. McGaw, & E. Care (eds.), *Assessment and Teaching of 21st Century Skills*. Dordrecht: Springer, pp. 17–66.

Bonney, R., Ballard, H., Jordan, R., McCallie, E., Phillips, T., Shirk, J., & Wilderman, C. C. (2009). *Public Participation in Scientific Research: Defining the Field and Assessing Its Potential for Informal Science Education*. A CAISE inquiry group report. Washington, D.C.: Center for Advancement of Informal Science Education (CAISE).

Bourdieu, P. (1986). Forms of capital. In J. Richardson (ed.), *Handbook of Theory and Research for the Sociology of Education*. New York, NY: Greenwood, pp. 241–258.

Braund, M., & Reiss, M. (2006). Validity and worth in the science curriculum: Learning school science outside the laboratory. *Curriculum Journal*, 17(3), 213–228.

Bunce, L. (2016). Appreciation of authenticity promotes curiosity: Implications for object-based learning in museums. *Journal of Museum Education*, 41(3), 230–239.

Dawson, E. (2014). Equity in informal science education: Developing an access and equity framework for science museums and science centres. *Studies in Science Education*, 50(2), 209–247.

Dawson, E., & Jensen, E. (2011). Towards a contextual turn in visitor studies: Evaluating visitor segmentation and identity-related motivations. *Visitor Studies*, 14(2), 127–140.

European Citizen Science Association (2015). *Ten Principles of Citizen Science*. Retrieved March 27, 2019 from https://ecsa.citizen-science.net/sites/default/files/ecsa_ten_principles_of_citizen_science.pdf.

Falk, J. (2009). *Identity and the Museum Visitor Experience*. Walnut Creek, CA: Left Coast Press.

Falk, J., & Dierking L. (2000). *Learning from Museums: Visitor Experiences and the Making of Meaning*. Walnut Creek, CA: Altamira Press.

Falk, J., & Dierking, L. (2013). *The Museum Experience Revisited*. London: Routledge Taylor and Francis Group.

Falk, J. H., & Dierking, L. D. (2002). *Lessons without Limit: How Free-choice Learning is Transforming Education*. Walnut Creek, CA: AltaMira Press.

Feinstein, N. W. (2017). Equity and the meaning of science learning: A defining challenge for science museums. *Science Education*, 101(4), 533–538.

Friedman, A. (ed.), (2008). *Framework for Evaluating Impacts of Informal Science Education Projects, Report from a National Science Foundation Workshop*. Arlington, VA: National Science Foundation. Retrieved March 27, 2019 from http://informalscience.org/documents/Eval_Framework.pdf.

Golumbic, Y. N., Baram-Tsabari, A., & Fishbain, B. (2016). Increased knowledge and scientific thinking following participation of school students in air-quality research. *Proceedings of Indoor Air 2016*. ISBN-13: 978-0-9846855-5-4, paper 832.

Gutwill, J. P., Hido, N., & Sindorf, L. (2015). Research to practice: Observing learning in tinkering activities. *Curator the Museum Journal*, 8(2), 151–168.

Hidi, S., & Renninger, K. A. (2006). The four-phase model of interest development. *Educational Psychologist*, 41(2), 111–127.

Hofstein, A., & Rosenfeld, S. (1996). Bridging the gap between formal and informal science learning. *Studies in Science Education*, 28, 87–112.

Hohenstein, J., & Moussouri, T. (2018). *Museum Learning: Theory and Research as Tools for Enhancing Practice*. Abingdon: Routledge.

Hooper-Greenhill, E., Dodd, J., Moussouri, T., Jones, C., Pickford, C., Herman, C., Morrison, M., Vincent, J., & Toon, R. (2003). *Measuring the Outcomes and*

Impact of Learning in Museums, Archives and Libraries, the Learning Impact Research Project End of Project Paper. Leicester: Research Centre for Museums and Galleries.

Jensen, E. (2014). Evaluating children's conservation biology learning at the zoo. *Conservation Biology, 28*(4), 1004–1011. doi: 10.1111/cobi.12263.

Jensen, E., & Buckley, N. (2014). Why people attend science festivals: Interests, motivations and self-reported benefits of public engagement with research. *Public Understanding of Science, 23*(5), 557–573. doi: 10.1177/0963662512458624.

Jensen, E., & Holliman, R. (2016). Norms and values in UK science engagement practice. *International Journal of Science Education — Part B: Communication and Public Engagement, 6*(1), 68–88. doi: 10.1080/21548455.2014.995743.

Jensen, E., & Wright, D. (2015). Critical response to Archer *et al.* (2015) "Science Capital": A conceptual, methodological, and empirical argument for extending Bourdieusian notions of capital beyond the arts. *Science Education, 99*(6), 1143–1146.

Kennedy, E. B., Jensen, E. A., & Verbeke, M. (2017). Preaching to the scientifically converted: Evaluating inclusivity in science festival audiences. *International Journal of Science Education Part B: Communication & Engagement, 8*(1), 14–21.

Land-Zandstra, A. M., Devilee, J. L. A., Snik, F., Buurmeijer, F., & Van den Broek, J. M. (2016). Citizen science on a smartphone: Participants' motivations and learning. *Public Understanding of Science, 25*(1), 45–60.

Lemke, J. L. (2000). Across the scales of time: Artifacts, activities, and meanings in ecosocial systems. *Mind, Culture, and Activity, 7*(4), 273–290.

Maslow, A. H. (1943). A theory of human motivation. *Psychological Review, 50*(4), 370–396. Retrieved March 27, 2019 from http://dx.doi.org/10.1037/h0054346.

McComas, W. F. (2006). Science teaching beyond the classroom. *Science Teacher, 73*(1), 26–30.

Moss, A., Jensen, E., & Gusset, M. (2017). Probing the link between biodiversity-related knowledge and self-reported pro-conservation behavior in a global survey of zoo visitors. *Conservation Letters, 10*(1), 33–40. doi: 10.1111/conl.12233.

National Research Council (2009). *Learning Science in Informal Environments: People, Places, and Pursuits*. Washington, DC: The National Academies Press.

Packer, J., & Ballantyne, R. (2002) Motivational factors and the visitor experience; a comparison of three sites. *Curator: The Museum Journal, 45*(3), 183–198. doi.ORG/10.1111/J.2151-6952.2002.TB00055.X.

Patrick, P. G. (2017). *Preparing Informal Science Educators.* Basel: Springer International Publishing.

Pedaste, M., Mäeots, M., Leijen, Ä., & Sarapuu, S. (2012). Improving students' inquiry skills through reflection and self-regulation scaffolds. *Technology, Instruction, Cognition and Learning, 9,* 81–95.

Pekarik, A. J., Doering, Z. D., & Karns, D. A. (1999). Exploring satisfying experiences in museums. *Curator: The Museum Journal, 42*(2), 152–173.

Petrich M., Wilkinson, K., & Bevan, B. (2013). It looks like fun, but are they learning? In M. Honey & D. E. Kanter (eds.), *Design, Make, Play: Growing the Next Generation of STEM Innovators.* New York, NY: Routledge.

Piaget, J. (1936). *Origins of Intelligence in the Child.* London: Routledge & Kegan Paul.

Simon, N. (2016). *The Art of Relevance.* Santa Cruz, CA: Museum 2.0.

Sjøberg, S., & Schreiner, C. (2010). *The ROSE Project: An Overview and Key Findings.* Retrieved March 27, 2019 from http://roseproject.no/network/countries/norway/eng/nor-Sjoberg-Schreiner-overview-2010.pdf.

Spitzberg, B. H. (1983). Communication competence as knowledge, skill, and impression. *Communication Education, 32*(3), 323–329. doi: 10.1080/03634528309378550.

Stocklmayer, S., Rennie L., & Gilbert J. (2010). The roles of the formal and informal sectors in the provision of effective science communication. *Studies in Science Education, 46*(1), 1–44.

Van Aalderen-Smeets, S. I., Walma van der Molen, J., & Asma, L. J. F. (2011). Primary teachers' attitudes toward science: A new theoretical framework. *Science Education, 96*(1), 158–182.

Van Gerven, D. J. J., Land-Zandstra, A. M., & Damsma, W. (2018). Authenticity matters: Children look beyond appearances in their appreciation of museum objects. *International Journal of Science Education, Part B, 8*(4), 325–339.

Vossoughi, S., & Bevan, B. (2014). Making and tinkering: A review of the literature. In *National Research Council Committee on Out of School Time STEM.* Washington, DC: National Research Council, pp. 1–55.

Vygotsky, L. S. (1978). *Mind in Society: The Development of Higher Psychological Processes.* Cambridge, MA: Harvard University Press.

Vygotsky, L., & Luria, A. (1994). Tool and symbol in child development. In J. Valsiner & R. van der Veer (eds.), *The Vygotsky Reader*. Oxford: Blackwell, pp. 99–172.

Wellington, J. (1990). Formal and informal learning in science: The role of the interactive science centers. *Physics Education*, 25, 247–252.

Chapter 6

Science Journalism

Mark Bos and Frank Nuijens

6.1 Introduction

When considering science journalism, newspapers and magazines quickly come to mind. But, TV and radio, as well as interpersonal networks, are also important sources of information about science for many people. Although not all science communication necessitates science journalists, much of the visible information about science in these aforementioned examples are provided by these professionals.

In a time where science and technology, and especially communication technology, is changing rapidly, science journalism is important. The manner in which people come into contact with science issues, be it through personal experience or communication, social or mass media, influences how they think about these issues. Therefore, the way people come into contact with issues such as climate change, genetic modification, and vaccination is important to consider (Hargreaves, Lewis & Speers, 2003) and getting high-quality information about science and technology to the public is vital (Dunwoody, 2014; Korthagen, 2016).

But, more is at play than just sending information. Today's media make it possible for all to partake in discussions and debates. Moreover, they make many things more transparent, such as the interactions between scientists and scientists, be they negative or positive. Also, they allow for easy publishing of science information and many scientists have picked up the pen themselves.

Digital communication enables people to deliver information via various routes. Today, people are less dependent on mass media for information about science and less dependent on science journalists to signal scientific breakthroughs. They can search the Internet for information about science issues they are interested in, they can read blogs and view vlogs by opinion leaders worth following, and, should they want to, they can post information about these issues themselves. The rise of the Internet and social media has dramatically increased both the amount of information available and the platforms where this information can be found. Somewhat paradoxically, this results in high quantities of information readily available, but the variety in platforms diminishes the chances that an individual is exposed to a particular chunk of information.

Thus, these are times in which the profession of science journalism is undergoing change, in which scientific developments are moving fast and deserve journalistic scrutiny and proper explanation, and in which media afford scientists with ample opportunity to interact with the publics. This chapter will describe different forms of science journalism and roles of science journalists and highlight some of the differences between general and science journalism, current trends, and developments in the context of the changing media landscape.

6.2 Science Journalism

As shown in the introduction, many authors (e.g. see Bauer & Bucchi, 2007) seem to concur that science journalism originated when scientists wanted to share their knowledge as widely as possible; this was followed by a period in which scientists increasingly moved away from direct contact with the publics, leaving a gap that was filled by science journalists (Dunwoody, 2014).

Fahy & Nisbet (2011), referring to, among others, Lewenstein (1992), Nelkin (1995), Rensberger (2009), and Trench (2009), also provide a good historical overview — indicating how the pendulum swung from a predominance of translation of novel developments in the 1900s to a more persuasive role in the 1930s and 1940s, to a more critical approach in the 1960s, or focus on the environment in the 1970s, and a *cheerleader* style in the 1980s, returning to a *watchdog* style in the 1990s. They state that the current digital age is typified by scientists self-publishing via blogs, social media, and personal websites. In a way, this most recent development can be seen as a shift back to those early days in the sense that new media have provided scientists with new means to share their knowledge as widely as possible.

Science journalism is a type of journalism that primarily deals with scientific achievements and breakthroughs, the scientific process itself, scientists' quests, and difficulties in solving problems (Angler, 2017). It includes all journalism related to science in the broadest sense of the word, including, but not limited to science as a process, scientific findings, science institutions (or individuals), and encompassing all types of science, including humanities and natural sciences (Korthagen, 2016; Wormer, 2009). Science and journalism are not alien cultures. They are built on the same foundation — the belief that conclusions require evidence, that the evidence should be open to everyone, and that everything is subject to question (*Nature*, 2009, p. 1033).

6.3 Forms of Science Journalism

According to various authors, science journalism can have different forms: informative, explanatory, and investigative, or mixtures of these three (e.g. Duits & Pleijter, 2016; Dunwoody, 2014; Peters, 2013).

Informative: The products of informative science journalism aim to inform the public in a primarily descriptive manner. Informative journalism focuses on bringing the news as it happens in order for the public to be informed about what happens in the world they live in. For example, informing a general public that CERN has discovered the Higgs boson particle.

Explanatory: Fahy & Nisbet (2011), among others, argue that the public should be informed not only about science but also about specific aspects of science, such as the scientific method(s) and its limitations. Indeed, some science journalists do not just inform but also explain things about science. Science journalism often has to explain complex issues. Consequently, large parts of science journalism articles will be explanatory, rather than solely informative. For example, explanations about what the Higgs boson particle is exactly. Explanatory journalism focuses on explaining and interpreting complex events and phenomena and journalists report about these issues in their social, political, and cultural contexts (Forde, 2007).

Investigative: In Brown (2014, pp. 829, 831), Evan Hansen, previously editor at Wired.com and currently senior editor at Medium.com, is cited, indicating that informative and explanatory science journalism is not enough, emphasizing the need for investigative journalism as well. More so than the other two forms of journalism, investigative journalism emphasizes the importance of criticasters; of people who take the effort to look beyond what is being said and check, and double check if what scientists claim is valid. In this role, science journalists become the so-called *watchdogs* of the society. This watchdog function of journalists is perhaps, traditionally, the most important role and is not new (see for example Fahy & Nisbet, 2011; Murcott & Williams, 2013; Rensberger, 2009). Rather than solely bringing the news and explaining the difficult aspects of that news, investigative science journalism checks whether there may be a conflict of interest (e.g. in the vaccination debate, where one of the issues is the degree to which science and pharmacy are intertwined) or whether conclusions are based on sound research (e.g. fraud research on the link between vaccination and autism) (Guenther *et al.*, 2019).

An example of investigative journalism is the blog — The Retraction Watch — which not only reports on research papers that are retracted from scientific journals but also tries to uncover why the research was retracted.

6.4 Roles of Science Journalists

In addition to functions of science journalism, science journalists may take on different roles. According to Fahy & Nisbet (2011), science journalists are performing a wide plurality of roles. Science journalists can have varied roles such as conduits and explainers, curators of information, civic educators, public intellectuals, agenda-setters, watchdogs, or conveners (pp. 786–789).

The agenda-setter identifies and calls attention to important areas of research, trends, and issues. The civic educator informs non-specialist audiences about the methods, aims, limits, and risks of scientific work. The conduits explain or translate scientific information in their reporting from experts to non-specialist publics. All three roles serve the informative function primarily. The public intellectual synthesizes a range of complex information about science and its social implications, which typifies the explanatory function. Finally, in relation to the investigative function, the investigative reporter carries out in-depth journalistic investigations into scientific topics. The watchdog holds scientists, scientific institutions, industry, and policy-oriented organizations up to scrutiny. The curator gathers science-related news, opinions, and commentary, presenting it in a structured format, with some evaluation. The convener connects and brings together scientists and various non-specialist publics to discuss science-related issues in public, either online or physically.

For an example, see Brüggemann (2017), who discusses how the roles of journalists who cover climate change has diversified over time. He finds that journalistic roles 'have evolved from an objective to a more interpretive form of journalism', where journalists give context to contrarian voices, for instance, 'by pointing to their lack of scientific credentials'.

6.5 News Criteria and Story Construction

In a broader perspective, including journalism and news in general, news and, thus, journalism should provide people with the information they need to act autonomously (Kovach & Rosenstiel, 2007). Whether a story is considered to be news depends on many factors. One factor is the media outlet. If the outlet is considered to deliver news, as for example a newspaper does, then the content presented is generally considered to be news. The prominence of content in such outlets is generally determined by news values or news criteria.

Galtung & Ruge (1965) were one of the first to identify these news criteria. They described 12 factors that are frequently used as defining the newsworthiness of a story. Based on their findings, they argued that the likelihood that something appears in the media increases with the number of factors satisfied. Of course, news values may differ across countries as they are culturally dependent and, at least to some extent, whether something becomes news is also dependent on arbitrary factors such as luck, convenience and serendipity (Harcup & O'Neill, 2016).

Harcup & O'Neill (2016) updated the list of Galtung and Ruge, affording for new outlets such as social media and including not only news about three international crises that was the basis of the news values of Galtung and Ruge but all journalistic content in the media outlet. Examples are factors relating to timing, significance, proximity, prominence, and human Interest (for a more elaborate overview, see Box 6.1). News factors are not only relevant for general news but also for science news. For instance, the factor 'conflict' from Box 6.1 can be found in stories about scientific controversies, 'surprise' in articles about scientific discoveries, and 'follow-up' in reporting on climate change.

Box 6.1: News factors as defined by Harcup & O'Neill (2016) in their paper, *What is news? News values revisited (again)*

1. *Exclusivity:* Stories generated by, or available first to, the news organization as a result of interviews, letters, investigations, surveys, polls, and so on. Exclusivity can be problematic in science journalism because of embargoes (see also Section 6.6).
2. *Bad news:* Stories with particularly negative overtones such as death, injury, defeat, and loss (of a job, for example).
3. *Conflict:* Stories concerning conflict such as (scientific) controversies, arguments, splits, strikes, fights, insurrections, and warfare.
4. *Surprise:* Stories that have an element of surprise, contrast, and/or the unusual about them, like an unexpected scientific discovery.
5. *Audio-visuals:* Stories that have arresting photographs, video, audio, and/or which can be illustrated with infographics.
6. *Share-ability:* Stories that are thought likely to generate sharing and comments via Facebook, Twitter, and other forms of social media.
7. *Entertainment:* Soft stories concerning sex, show business, sports, lighter human interest, animals, or offering opportunities for humorous treatment, witty headlines, or lists with scientific factoids.
8. *Drama:* Stories concerning an unfolding drama such as escapes, accidents (like the accident in the nuclear facility of Fukushima), searches, sieges, rescues, battles, or court cases.
9. *Follow-up:* Stories about (scientific) subjects already in the news.
10. *The power elite:* Stories concerning powerful individuals, organizations, institutions, or corporations.
11. *Relevance:* Stories about groups or nations perceived to be influential with, or culturally or historically familiar to, the audience.
12. *Magnitude:* Stories perceived as sufficiently significant in the large numbers of people involved or in potential impact or involving a degree of extreme behavior or extreme occurrence.

Box 6.1: (*Continued*)

13. *Celebrity.* Stories concerning people who are already famous, also famous scientists like Einstein and Nobel laureates.
14. *Good news*: Stories with particularly positive overtones such as recoveries, breakthroughs, cures, wins, and celebrations.
15. *News organization's agenda*: Stories that set or fit the news organization's own agenda, whether ideological, commercial, or as part of a specific campaign.

Some of the news factors in Box 6.1, such as factors 1, 3, and 9, clearly relate to what the story is about, to its content. Others relate more to other aspects, such as whether the story is easy to share, which seems more media dependent (for example, factor 6) and organization dependent (for example, factor 15). Especially this last factor clearly shows that, regardless of the event, the journalist and organization have an important role in news creation.

In this sense, news and news values can be perceived as reflections of organizational, sociological, and cultural norms, combined with economic factors, perhaps more so than as something that serves a public and its information need (Weaver *et al.*, 2009). This also explains why the list is not constant and needs to be updated regularly to afford for organizational, sociological, and cultural trends.

Journalists make use of news criteria when selecting and writing their stories to increase their chances of publication. Journalists want their complete news items to be read, seen, and heard. To achieve that, they make use of specific techniques to construct their stories. For example, Kruyzen (2007) and others (see also Bednarek & Caple, 2012) refer to intensifying particular aspects of an event. Montgomery (2007) points out unambiguous reporting. Authors also identified some universal practices applied in journalism in general that are applicable in science journalism as well. An example is the *editorial coloring Box* (Kruyzen, 2007). Kruyzen referred to the techniques used by journalists and editorial teams as a coloring Box because they like to paint their stories in bright colors and, by doing this, attract attention and communicate their message. These elements are as follows: focusing (on just one aspect of the story), simplifying (for a better understanding), polarizing (emphasizing conflicting interests), exaggerating (embroidering on reality), intensifying (building tension), illustrating (clarifying complex information), and personifying (emphasizing human aspects).

6.6 Differences between General News and Science News

Although science news follows the above-described news criteria and logic, some important differences between news and science news exist, such as how science journalists should deal with balance, embargoes, and (un)certainty.

(*False*) *Balance*: Balance refers to the standard principle in journalism of hearing both sides of an argument. This frequently results in giving both sides of the story equal space in the story. However, many authors argue that science journalism requires a different sort of balance than general journalism. In science, sometimes the scientific evidence of one side of the argument is much stronger than that of the other side. In such instances, it is misleading to give both sides equal space in the story.

Clarke (2008) addressed this issue in relation to the autism-vaccine controversy, referring also to Antilla (2005), Boykoff & Boykoff (2004), and Ryan (2001). Clarke (2008) indicates that balance and accuracy, two norms that traditionally govern media coverage of controversial issues, can sometimes be at odds in science journalism. On the one hand, balance demands that journalists present all sides of an issue. On the other hand, accuracy demands scrutiny of details and verification of facts to avoid errors, which seems to point out that a journalist should convey to the public which perspective has the most supportive evidence.

A much-debated example is the issue of climate change. Van der Sluijs *et al.* (2010) indicate that it is not so much the question if there is a climate problem, but rather, what factors influence climate change. Many believe, concerning this issue, that too many media give too much attention to skeptics who do not always bring in scientific arguments. An accurate representation of the scientific certainty about climate change would give 97% of the space to those who are in the consensus while 3% would be given to those who disagree.

The BBC addressed this issue when the BBC Trust commissioned a report to review the impartiality and accuracy of the BBC's coverage of science (BBC Trust, 2011). One of its conclusions is that, 'programme makers must make a distinction between well-established fact and opinion in science coverage and ensure the distinction is clear to the audience' (p. 3).

Therefore, many authors (e.g. Boykoff & Boykoff, 2004; Clarke, 2008; Corbett & Durfree, 2004; Oreskes & Conway, 2010; Wahl-Jorgensen *et al.*, 2017) have indicated that, in science journalism, journalists may need to act differently when using balance in science reporting. When a lot of scientific evidence for a particular view is present, this should be reflected in the science reporting. The traditional *pros and cons* framing that is typical for regular journalism is too simplistic in these cases (Korthagen, 2016).

David Robert Grimes, a physicist and cancer researcher at Oxford University, summarizes the dangers of false balance in an opinion article in *The Guardian* in 2016 (Tuesday, 8 November) as follows: 'Media outlets have an important role to play in conveying vital information and view-points, pushing a standard for fact-checking and quality control that may be lacking in more fragmented modern media. But engaging in false balance undermines this strength and risks giving debunked or dangerous fringe views an air of legitimacy and the oxygen of publicity — and ultimately, such sophism leaves us all more divided and less informed.'

Clarke (2008, p. 79) argues that on one hand, balance may be a function of the quantity of information presented, in which a journalist highlights all relevant viewpoints regardless of how well-known or influential they may be. On the other hand, balance may be a function of quality, in which one identifies the two most influential perspectives, presents them in a point-counterpoint format, and affords both relatively equal attention. How journalists should best adhere to the balance norm while also addressing the public's right to be informed, stakeholders' rights to have their perspectives heard in the public arena, and their own responsibility to convey truthful information in their reporting' is still the topic of much debate.

Dealing with embargoes: Something that should, or, perhaps better put, could aid journalists' scrutiny of the balance issue, is the embargo system. Many scientific journals, such as *Nature, Science,* and the *Proceedings of the*

National Academy of Sciences (*PNAS*), impose embargoes on journalists wanting to report on papers published in those journals. This means that journalists are given a list of the most noteworthy publications that are due to appear, usually a week in advance. Journalists who want access to this list have to agree to postpone the publication of a story on the embargoed publications until the embargo expires. This system has several advantages and also disadvantages.

The most important advantage for journalists is that they get ample time to prepare an in-depth article. After all, journalists do not have to compete who gets the story first. They all have to wait until the scientific journal's embargo lifts. A downside of this is that the embargo system may make journalists lazy. Kiernan (2006) argues that the embargo system diverts attention from pursuing original stories to other issues because editors may want them to cover the stories that everyone else also covers.

This benefit–downside trade-off was also recognized by Marshall (1998, p. 860), who stated: 'Journals get maximum publicity, journalists get time to report complex stories, and scientists get more widespread and more accurate public exposure for their work." He warned that embargoes can impede the dissemination of scientific results: "Science is supposed to progress through rapid communication of results among scientists, but the embargo system can erect barriers to this exchange of information.'

(*Un*)*Certainty*: Science journalists also have to handle the aspect of certainty differently. Science is all about uncertainty, rather than about certainty, which is why much can be said about how journalists deal with scientific uncertainty (e.g. Stocking, 1999; Brossard *et al.*, 2019). Despite increased acknowledgement of this aspect of inherent uncertainty in science news, much of the information is still presented in the form of universal truths in news outlets (Dijstelbloem *et al.*, 2013). One reason is, perhaps, that journalists are generally trained to provide clear and decisive news and have to adhere to the frame of their medium and refrain from the temptation to see connections where there are none or draw false conclusions from reported results.

Not all journalists seem to understand the meaning of p-values in statistics or the difference between a correlation and causality. Spurious correlations are frequently reported in science news. One example is research showing a positive correlation between light drinks and weight, which was reported as a causal relation. The Dutch Nutrition Centre indicated that research has indeed shown a relation between the two, but this relation was explained by the fact that people who are overweight are more likely to drink light products than people whose weight is normal.

The completion of a satellite built by NASA is a certainty. But, much else in science will always include an amount of uncertainty — this is inherent to the scientific method and philosophy. Science journalists should be aware that sometimes scientists themselves will exaggerate in order to attract media attention for their research or provide misleading information. Haneef *et al.* (2015), for example, indicated that scientists themselves use specific reporting strategies (like spinning their story, where they sell a message that is biased in favor of their own position) to convince readers, among whom are journalists, that their research is important, more important than warranted by the results.

6.7 The Changing Media Landscape

When writing about science journalism, news, and trends, one aspect stands out: media, or, perhaps better put, the changing media landscape and the transition from offline to online. In order to survive, many science information outlets have adopted cross-media approaches, using multiple media to convey supplementing information, making maximum use of medium-specific merits. *Scientific American*, for example, produces over 3,000 items a year, including articles, news stories, videos, podcasts, slideshows (Brown, 2014), and does so in magazines, on their website, as well as via Instagram, YouTube, Twitter, and Facebook.

However, in cross-media approaches, traditional media are still important. Television continues to play an important role in science communication, although perhaps science *news* on TV remains scarce (Mellor, Webster & Bell, 2011). TV is an important medium, primarily because of its reach, which is still high. Also important, although not unique of course, is its capability to combine (moving) images with sound. In most instances, the budgets for television are much larger than for newspapers, blogs, or vlogs. Despite many benefits, downsides also exist. Science journalism on TV requires good visual and auditory materials, as well as a proficient speaker. Also, the medium is less suitable to communicate complex information. Much of the research shows that viewers remember little of what they have seen. And, it is expensive.

Apart from television, radio is also a traditional medium. It is still a popular medium, in particular in developing countries. Radio is a very direct medium. This means that, in science journalism, scientists are often able to tell their stories themselves. However, similar to TV, an important downside is, of course, that radio solely depends on audio cues, so, again, it is very dependent on speaker skills. Another thing to realize is that, for most people listening to the radio is a secondary activity, so much of the information may be missed entirely or is received only subliminally.

However, radio has benefited from recent trends and technological developments. Many radio shows supplement their programming with live video feeds, podcasts, and websites. In this format, radio shows can be made searchable and people can listen to them in their own time rather than solely at the time it is broadcast live.

Older than both TV and radio is, of course, the newspaper. Newspapers generally have more room to elaborate than TV and radio. Journalists can emphasize subtle distinctions that are sometimes omitted in TV and radio productions. In addition, newspapers still enjoy public trust. Important downsides are that printed newspapers nowadays are considered slow when compared to digital news outlets. This is precisely why print newspapers generally supplement this medium with online content. Another downside is that people are selective readers, and it is difficult to track what articles are being read for print newspapers, which limits the possibilities for editors to follow-up on much read articles. Finally, the content of printed newspapers cannot be easily shared. This is a highly important and generally valued aspect of online media. Despite the fact that the numbers of printed newspapers drop, online numbers are on the rise, albeit, online numbers are not as high as printed numbers once were (Korthagen, 2016).

In contrast to print and other traditional media, online possibilities may seem endless. One of the most important benefits is that online content is easily searchable. Another benefit is that one can link one's content with that of others, for example, to provide links to original materials. Also, the online space is never full. No spatial limit on texts, images, and sounds is present, as there is in print newspapers. In this context, online is also almost by definition multimedia. It can facilitate *traffic*, thereby bringing more users to read/view the content.

However, online content has several disadvantages. Search engines are frequently used by people to search for science information (e.g. Bos, Koolstra & Willems, 2009; Brossard & Scheufele, 2013), but these search engines use pre-programmed self-learning algorithms. The search is limited to the words known and used by the user. If these words are not used in science, then the findings may be biased to non-science sources. Also, the search engine makes use of earlier searches and visited pages. This creates a risk of an ever-increasing self-confirmatory bias.

Because the online is never full, one must be careful for (cognitive) overload. Too lengthy texts on websites are good examples. Also, the amount of information online is huge, which means high competition for attention as well, not only across media but also within the medium itself. Finally, because anybody and everybody can post information online, a lot of discussion has come up about the value or trustworthiness of online information. As such, it is sometimes difficult for people to assess the quality of the information delivered.

In this regard, it is also important to realize that each social medium has its own specific characteristics. Solis (2008, 2017) developed the so-called conversation prism to show exactly what social media, or, perhaps better put in this context, platforms exist and what they are most frequently used for. This shows, for example, that Facebook, LinkedIn, and Twitter are not interchangeable.

6.8 Trends in Science Journalism

Trends in science journalism occur at different levels. Some of the more important trends are as follows: churnalism, storytelling, fake news, constructive journalism, and slow journalism.

Churnalism: Churnalism refers to the recycling of pre-packaged public relations and press agency copy as news (Davies, 2008). It is a phenomenon that has become quite big. In some instances, churnalism makes up almost half of the news stories published in quality press (Lewis, Williams & Franklin, 2008). In the UK, the Science Media Centre, which aims to act as an 'independent press office' for science, has been criticized for making it easier for journalists to copy/paste press releases (Davies & Horst, 2016, p. 81; see Ogden, 2015 for a discussion on this critique). This type of journalism is prone to error when journalists neglect to read all information available. Without proper scrutiny, the journalist cannot determine whether the press release, or, for that matter, the scientist, comes to a valid conclusion. The Media Standards Trust has launched a website enabling journalists to compare press releases to national newspaper articles (at the time of writing, the website was offline due to problems with funding). *The Guardian* wrote a piece about it in 2011 providing examples of newspapers copying up to 98% of the text directly from a press release by the Benenden Healthcare Society on 'British women spend more money on their looks than their health'.

Storytelling: Storytelling uses narrative journalistic stories. These tend to be longer pieces of work and include elements of prose writing, rather than the more traditional objective factual writing. Narrative non-fiction storytelling, as a genre, is framed as moving, essential, and as high-quality journalism (Van Krieken & Sanders, 2017). But, it comes at a risk because the journalistic principles of objectivity and factuality are relegated to the background in order to facilitate the storytelling elements. *The New Yorker* is an example of a magazine that publishes narrative journalistic stories and *The Guardian* also publishes science news in this format; see the article by Zoë Corbyn, 'Want to Live Forever? Flush Out your Zombie Cells' (Corbyn, 2018). Storytelling can be a strong driver for engagement in science communication, as Martinez-Conde & Macknik (2017, p. 8129) pose: 'Reaching a general audience while communicating scientific content is perhaps as much an art as a science, and successful art engenders emotion. Identifying and developing such emotional connections in the public might be a powerful path to a gripping plot.'

Fake news: The issue of fake news has received ample attention recently, especially in relation to US elections (e.g. Allcott & Gentzkow, 2017; McNair, 2017). But, obviously, fake news is not restricted to news about politics and policy (see Scheufele & Krause, 2019, for an overview of how and why citizens become misinformed about science). Marcon, Murdoch & Caulfield (2017), for example, discussed the issue in relation to stem cell research. Fake news is news that favors distorted, decontextualized, or dubious information (Hunt, 2016). To some extent, fake news is purposive.

For example, creating click-bait for commercial purposes (McCoy, 2016). But, with the increased pressure on timeliness in the current era, journalists struggle to obtain credible information quickly (Brown, 2014), limiting the possibility to cross-check information, increasing the chance of fake news entering the arena. Fake news is not something new. The tobacco industry, for example, introduced fake news in the past to obscure the truth about tobacco smoke (Oreskes & Conwey, 2010).

Constructive journalism: As a reaction to the seemingly inherent negative bias of journalism, constructive journalism has come up. It is a form of journalism that involves applying positive psychological techniques to news processes and production in an effort to create productive and engaging coverage (McIntyre & Gyldensted, 2017). The Dutch medium *De Correspondent* is an example of an online platform focusing on constructive journalism. In Great Britain, the BBC also works on it. *The Constructive Journalism Project* is an organization giving workshops to journalism students and freelance journalists. *The Guardian* published an article about the issue by Laura Oliver (August 1, 2016) as well: 'And now for the good news: why the media are taking a positive outlook.'

Slow journalism: In turn, slow journalism is a reaction to the existing 24/7 mentality of today's world. As such, parallel to fast journalism, with its inherent risks, slow journalism is less focused on the now and more on the relevance of an issue (Le Masurier, 2015). *Delayed Gratification* is an example of a magazine focusing on this approach to news. Blanding (August 19, 2015) wrote an article about it as well: 'The value of slow journalism in the age of instant information'. Another interesting example is the website *The Conversation*, where journalists and scientists work together to offer scientific knowledge, insights, and opinions to news events. The platform started in Australia and has since expanded with dedicated sites in the UK, US, Africa, France, and Canada.

6.9 Science Journalism: From Sender to Receiver

Overall, the traditional boundaries between sources, journalists, and audiences are fading (Brown, 2014; Dunwoody, 2014). As a result, the landscape of science journalism is changing. Rather than news per day (traditional newspapers), news per hour (traditional TV), or news per half hour (traditional radio), people nowadays expect minute-by-minute updates (social media). Timeliness has always been, and will always be, an important aspect of science journalism. But, as Brown (2014) puts it, where in the past journalists would hear breaking news by word of mouth or over the radio, nowadays they learn the latest news via social media, such as Twitter. Of course, Twitter was big in 2014, but has been surpassed by platforms such as WhatsApp and Instagram, which, in turn, will be surpassed by others in the near future. Keeping track of these developments has become the job of the science journalist as well.

In order to facilitate this minute-by-minute news feed, it has become difficult to uphold some of the traditional journalistic virtues, such as critically assessing the quality of the information and to write extensive explanations in order to provide context. Zimmer (in Brown, 2014) indicated that science journalism has improved a lot and that while long articles were one time shunned, currently they are increasingly the norm. This is at least in part, due to technological developments (e.g. better monitors to allow for easier online reading), increased online options (e.g. better fonts), and professional experience (e.g. better designs).

Finally, the transition from offline to online allows people to network via the web (Brown, 2014). The web, especially social network sites, enables people to find like-minded others, share ideas, and respond to content. More people than ever before will be able to read or watch journalists' work, but, conversely, precisely because so many people are able to read or watch this work online and, as well as because so many different ways to attain information are available, it has become difficult to monetize the efforts made (Brown, citing Fagin, 2014, p. 827).

From a commercial perspective, journalism is under pressure as traditional business models have become unsustainable (Korthagen, 2016). As a result, science journalism, in the past dominated by a limited number of professional organizations with full-time staff, has now become something more diffuse, a 'system where content is delivered by people being paid per job, or by bloggers and other contributors being paid very little, if anything at all' (Brown, 2014, p. 827).

As previously indicated, nowadays it is more about who delivers the content than it is about where the content appears. People have become less loyal to particular news outlets and have become more likely to follow individual scientists or journalists, regardless of where they publish their stories (Brown, 2014). This seems at least to be true for highly engaged audiences. Brown (2014) continues: 'the lines between sources, journalists and audiences are increasingly blurred by blogs and social networks, such that readers today often expect to be a part of the story' (p. 827).

Science journalism has moved from a predominantly sender-oriented perspective, in which science information trickles down from scientists, through science journalists, to various audiences, to a more receiver-oriented perspective, in which science information is shared between multiple actors, including science journalists. Within this trend, the role of citizens and their participation in the discussion about science issues, or even in the scientific process itself, is considered increasingly important. In this light, the changes in science journalism are related to changes in society in general.

Another change is the increased attention to the audience and increased room to include personal experiences in story development. Especially the inclusion of personal experiences in relation to science allows the public to reframe science; do people experience in real life what they are being told in the media? In a sense, it is a journalist vs public stress test (Secko *et al.*, 2011). New media have enabled audiences to make their own decisions about what is news, with user-generated content becoming more important in news production (Domingo *et al.*, 2008). However, this interaction between scientists, science journalists, and publics needs to be monitored and managed to be successful.

6.10 Journalism Influencing Audiences

As described above, new media have changed the way science is communicated, not only in science journalism but also in all related arenas. This influence extends to the way journalism determines what and how people think about issues (e.g. Peters, 2013). The theory of *agenda-setting* was first published in 1972 by McCombs and Shaw. This theory is still relevant in communication research as well as in science communication research (e.g. Vargo, Guo & Amazeen, 2018). In short, the agenda-setting theory states that when an issue receives attention in the mass media, when it is high on the news agenda, it will gain prominence on other agendas as well, such as, most notably, public and political agendas. The theory states not necessarily that media determine what people think, but what the topics are that people think about.

In recent years, this theory has been extended to afford for the cross-media, or, as Funk & McCombs (2017) put it, intermedia effects. This intermedia agenda-setting occurs when elite media influence smaller news agencies, which typically happens when journalists validate work by looking at their peers (Reese & Danielian, 1989; McCombs, 2018). Emerging media, such as blogs and vlogs, have been shown to become increasingly powerful in setting the agenda of other media (Vargo & Guo, 2017). An example is that trending topics on social media platforms are frequently picked up by news outlets these days.

Where agenda-setting theory predominantly looks at how media influence what people think about, *framing* theory looks at the manner in which information is presented, or framed, and influences how people think about it. For example, Nerlich (2015) highlights that especially metaphorical framing of science and scientists is frequently used when scientists are described as 'playing God' and linked to 'Frankenstein'. This is what Prince Charles from the UK did when he coined the term *Frankenstein food* in relation to genetically modified organisms. This type of framing obscures the way science really works, but seems almost inevitable. Framing determines how audiences interpret the information.

6.11 Audiences

In recent years we have seen an increase in the number of scientists that communicate directly with the public. In many of the publications referred to, these scientists are addressed and discussed in the context of science journalism, but generally referred to as science writers (e.g., Dunwoody, 2014), not science journalists. Similarly, Peters (2013) indicated that nearly all science organizations and many individual scientists have started to communicate with audiences themselves — indicating that the content they create and share competes with that provided by professional science journalists, but they are not labelled science journalists themselves. Brumfiel (2009) argues that future research should further investigate the effects of scientists-run vs journalists-run outlets. In particular, it should explore the distinction between science journalism and science communication in scientists-run outlets.

One aspect of science journalism remains somewhat underrepresented in this chapter: the audiences. Fahy & Nisbet (2011) indicate that in this news media landscape, especially highly motivated people can revel in available information about science. These audiences not only consume information but also contribute, recommend, share, and comment on news. However, science communication in general, and science journalism specifically, should not solely be for the interested and motivated few.

Also playing a role with regard to audiences is the intercultural aspect. Most of the literature on science communication, including that on science journalism, is from the US, UK, and Europe. But, science reporting and journalism are on the rise in the East as well (e.g. El-Awady, 2009). It is important to realize that across cultures, researchers and practitioners struggle not only with similar but also with very different issues. For example, accessibility of information can differ greatly in counties. El-Awady (2009) states that it is sometimes easier to find out what is happening halfway across the globe than what is happening locally. This indicates that challenging problems of government-owned institutions is not always possible in a culture. Other examples are language barriers. This influence is also visible in mass media. Ren *et al.* (2014), for example, showed clear differences between the UK and China about the way biomedical expertise is represented in public communication of health issues related to social controversy.

It is important to continuously keep track of the audience. When journalists monitor not only how the information comes through but also how it is received (processed) and used, science journalism may improve as a profession. It will allow for better targeting and tailoring of future communication. Online communication has made it easier to monitor what audiences do with the information available to them. It is now much easier to measure, for example, what news articles are read, shared, and discussed.

As discussed in the section on trends, the pace of communication has changed and audiences are increasingly becoming used to soundbites. In order to reach audiences via fast-paced outlets, science journalists as well as scientists need to be creative, speak the right language, and be able and willing to express oneself in one-liners (Bos & Van der Gaag, 2014).

6.12 Conclusion

To conclude, this chapter has shown that science journalism has a strong role to play in a society that is more and more knowledge driven, both in developed and developing countries. The terminology that journalists use to report on science issues (framing) and the quantity and quality of arguments in scientific debates that they present to the public (balance) determine how audiences will use scientific knowledge to make decisions in their everyday lives. To help untangle the science communication web, research should continue to investigate this triangle in which the actions of publics, scientists, and journalists continuously influence each other.

References

Allcott, H., & Gentzkow, M. (2017). Social media and fake news in the 2016 election. *Journal of Economic Perspectives, 31*(2), 211–236.

Angler, M. W. (2017). *Science Journalism: An Introduction*. London: Routledge.

Antilla, L. (2005). Climate of scepticism: US newspaper coverage of the science of climate change. *Global Environmental Change, 15*(4), 338–352.

Bauer, M. W., & Bucchi, M. (2007). *Journalism, Science and Society: Science Communication between News and Public Relations*. New York: Routledge.

BBC Trust (2011, July). BBC Trust Review of Impartiality and Accuracy of the BBC's Coverage of Science. With an independent assessment by Professor Steve Jones and content research from Imperial College London. London: BBC Trust.

Bednarek, M., & Caple, H. (2012). *News Discourse*. London: Bloomsbury academic.

Blanding, M. (2015, August 19). The value of slow journalism in the age of instant information. *Nieman Reports*.

Bos, M. J. W., Koolstra, C. M., & Willems, J. T. (2009). Adolescent responses toward a new technology: First associations, information seeking and affective responses to ecogenomics. *Public Understanding of Science, 18*(2), 243–253.

Bos, M. J. W., & Van der Gaag, B. (2014). Wetenschap in de media. In F. Van Dam, L. De Bakker, & A. M. Dijkstra (eds.), *Wetenschapscommunicatie, een kennisbasis*. Den Haag: Boom Lemma uitgevers, pp. 141–166.

Boykoff, M. T., & Boykoff, J. M. (2004). Balance as bias: Global warming and the US prestige press. *Global Environmental Change, 14*(2), 125–136.

Brossard, D., & Scheufele, D. A. (2013). Science, new media, and the public. *Science, 339*(6115), 40–41.

Brossard, D., Belluck, P., Gould, F., & Wirz, C. D. (2019, January 14). Promises and perils of gene drives: Navigating the communication of complex, post-normal science. *Proceedings of the National Academy of Sciences of the United States of America*. doi:10.1073/pnas.1805874115.

Brown, P. (2014). An explosion of alternatives: Considering the future of science journalism. *EMBO Reports, 15*(8). doi: 10.15252/embr.201439130.

Brüggemann, M. (2017). Shifting roles of science journalists covering climate change. *Oxford Research Encyclopedia of Climate Science*. Oxford: Oxford University Press USA. doi: 10.1093/acrefore/9780190228620.013.354.

Brumfiel, G. (2009). Science journalism: Supplanting the old media? *Nature News, 458*(7236), 274–277.

Clarke, C. E. (2008). A question of balance: The autism-vaccine controversy in the British and American elite press. *Science Communication, 30*(1), 77–107.

Corbett, J. B., & Durfree, J. L. (2004). Testing public (un)certainty of science: Media representations of global warming. *Science Communication, 26*, 129–151.

Corbyn, Z. (2018, October 6). Want to live for ever? Flush out your zombie cells. *The Guardian*. Retrieved from https://www.theguardian.com/science/2018/oct/06/race-to-kill-killer-zombie-cells-senescent-damaged-ageing-eliminate-research-mice-aubrey-de-grey.

Davies, N. (2008, February 4). Churnalism has taken place of what we should be doing: Telling the truth. *Press Gazette*.

Davies, S. R., & M. Horst (2016). *Science Communication: Culture, Identity and Citizenship*. London: Palgrave Macmillan.

Dijstelbloem, H., Huisman, F., Miedema, F., & Mijnhardt, W. (2013, September 9). Waarom de wetenschap niet werkt zoals het moet, en wat daar aan te doen is. *Science in Transition. Position paper*.

Domingo, D., Quandt, T., Heinonen, A., Paulussen, S., Singer, J. B., & Vujnovic, M. (2008). Participatory journalism practices in the media and beyond: An international comparative study of initiatives in online newspapers. *Journalism Practice, 2*(3), 326–342.

Duits, L., & A. Pleijter (2016). *"Uit onderzoek blijkt..." Een inventarisatie van wetenschapsjournalistiek in Nederlandse media*. Den Haag: Rathenau Instituut.

Dunwoody, S. (2014). Science journalism: Prospects in the digital age. In *Routledge Handbook of Public Communication of Science and Technology*. London: Routledge, pp. 43–55.

El-Awady, N. (2009). Science journalism: The Arab boom. *Nature, 459*(7250), 1057.

Fahy, D., & Nisbet, M. C. (2011). The science journalist online: Shifting roles and emerging practices. *Journalism, 12*(7), 778–793.

Forde, K. R. (2007). Discovering the explanatory report in American newspapers. *Journalism Practice, 1*(2), 227–244.

Funk, M. J., & McCombs, M. (2017). Strangers on a theoretical train: Inter-media agenda setting, community structure, and local news coverage. *Journalism Studies, 18*(7), 845–865.

Galtung, J., & Ruge, M. H. (1965). The structure of foreign news: The presentation of the Congo, Cuba and Cyprus crises in four Norwegian newspapers. *Journal of Peace Research, 2*(1), 64–90.

Grimes, D. R. (2016, November 8). Impartial journalism is laudable. But false balance is dangerous. *The Guardian*.

Guenther, L., Bischoff, J., Löwe, A., Marzinkowski, H., & Voigt, M. (2019). Scientific evidence and science journalism: Analysing the representation of (un) certainty in German print and online media. *Journalism Studies, 20*(1), 40–59.

Haneef, R., Lazarus, C., Ravaud, P., Yavchitz, A., & Boutron, I. (2015). Interpretation of results of studies evaluating an intervention highlighted in Google health news: A cross-sectional study of news. *PloS One, 10*(10). doi.org/10.1371/journal.pone.0140889

Harcup, T., & O'Neill, D. (2016). What is news? News values revisited (again). *Journalism Studies, 18*(12), 261–280.

Hargreaves, I., Lewis, J., & Speers, T. (2003). *Towards a Better Map: Science, the Public and the Media*. Swindon: Economic and Social Research Council.

Hunt, E. (2016, December 17). What is fake news? How to spot it and what you can do to stop it. *The Guardian*.

Kiernan, V. (2006). *Embargoed Science*. Chicago: University of Illinois Press.

Korthagen, I. A. (2016). *Wakers van de wetenschap. Over het belang en de functies van wetenschapsjournalistiek*. Den Haag: Rathenau Instituut.

Kovach, B., & Rosenstiel, T. (2007). *The Elements of Journalism. What Newspeople Should Know and the Public Should Expect*. New York: Three Rivers Press.

Kruyzen, H. (2007). *Journalisten bijten niet!* Zaltbommel: Haystack Uitgeverij.

Le Masurier, M. (2015). What is slow journalism? *Journalism Practice, 9*(2), 138–152.

Lewenstein, B. V. (1992). The meaning of 'public understanding of science' in the United States after World War II. *Public Understanding of Science, 1*(1), 45–68.

Lewis, J., Williams, A., & Franklin, B. A. (2008). Four rumours and an explanation: A political economic account of journalists' changing newsgathering and reporting practices. *Journalism Practice, 2*(1), 24–47.

Marcon, A. R., Murdoch, B., & Caulfield, T. (2017). Fake news portrayals of stem cells and stem cell research. *Regenerative Medicine, 12*(7), 765–775.

Marshall, E. (1998). Embargoes; Good, bad, or 'necessary evil'? *Science, 282*(5390), 860–867.

Martinez-Conde, S., & Macknik, S.L. (2017). Opinion: Finding the plot in science storytelling in hopes of enhancing science communication. *Proceedings of the National Academy of Sciences of the United States of America, 114*(31), 8127–8129.

McCombs, M. (2018). *Setting the Agenda: Mass Media and Public Opinion*. Malden: John Wiley & Sons.

McCombs, M. E., & Shaw, D. L. (1972). The agenda-setting function of mass media. *Public Opinion Quarterly, 36*(2), 176–187.

McCoy, T. (2016, November 20). For the 'new yellow journalists', opportunity comes in clicks and bucks. *The Washington Post*.

McIntyre, K., & Gyldensted, C. (2017). Constructive journalism: An introduction and practical guide for applying positive psychology techniques to news production. *The Journal of Media Innovations, 4*(2), 20–34.

McNair, B. (2017). *Fake News: Falsehood, Fabrication and Fantasy in Journalism*. London: Routledge.

Mellor, F., Webster, S., & Bell, A. R. (2011). Content analysis of the BBC's science coverage. Science Communication Group, Imperial College London. Retrieved March 29, 2019 from http://downloads.bbc.co.uk/bbctrust/assets/files/pdf/our_work/science_impartiality/appendix_a.pdf.

Montgomery, M. (2007). *The Discourse of Broadcast News: A Linguistic Approach*. London: Routledge.

Murcott, T. H., & Williams, A. (2013). The challenges for science journalism in the UK. *Progress in Physical Geography, 37*(2), 152–160.

Nature. (2009). Cheerleader or watchdog. *Nature, 459*, 1033.

Nelkin, D. (1995). *Selling Science: How the Press Covers Science and Technology*. New York, NY: W.H. Freeman and Company.

Nerlich, B. (2015). Metaphors in science and society: The case of climate science and climate scientists. *Language and Semiotic Studies, 1*(2), 1–15.

Ogden, E. L. (2015). Minding the media gap. *BioScience, 65*(3), 231–236.

Oreskes, N., & Conway, E. M. (2010). Defeating the merchants of doubt. *Nature, 465*(7299), 686.

Peters, H. P. (2013). Gap between science and media revisited: Scientists as public communicators. *Proceedings of the National Academy of Sciences of the United States of America, 110*(Supplement 3), 14102–14109.

Reese, S. D., & Danielian, L. H. (1989). Intermedia influence and the drug issue: Converging on Cocaine. In P. J. Shoemaker (ed.), *Communication Campaigns about Drugs: Government, Media, and the Public*. Hillsdale, NJ: Lawrence Erlbaum Associates, pp. 29–46.

Ren, J., Peters, H. P., Allgaier, J., & Lo, Y. Y. (2014). Similar challenges but different responses: Media coverage of measles vaccination in the UK and China. *Public Understanding of Science, 23*(4), 366–375.

Rensberger, B. (2009). Science journalism: Too close for comfort. *Nature, 459*(7250), 1055.

Ryan, M. (2001). Journalistic ethics, objectivity, existential journalism, standpoint epistemology, and public journalism. *Journal of Mass Media Ethics, 16*(1), 3–22.

Scheufele, D. A., & Krause, N. M. (2019, January 14) Science audiences, misinformation, and fake news. *Proceedings of the National Academy of Sciences of the United States of America*. doi: 10.1073/pnas.1805871115.

Secko, D. M., Tlalka, S., Dunlop, M., Kingdon, A., & Amend, E. (2011). The unfinished science story: Journalist–audience interactions from the globe and mail's online health and science sections. *Journalism, 12*(7), 814–831.

Solis, B. (2008, 2017). *The conversation prism*. Version 5. Retrieved March 26, 2019 from https://conversationprism.com/.

Stocking, S. H. (1999). How journalists deal with scientific uncertainty. In S. M. Friedman, S. Dunwoody, & C. L. Rogers (eds.), *Communicating Uncertainty: Media Coverage of New and Controversial Science*. Mahwah, NJ: Lawrence Erlbaum Associates, pp. 23–41.

Trench, B. (2009). Science reporting in the electronic embrace of the internet. In R. Holliman, E. Whitelegg, E. Scanlon, S. Smidt, & J. Thomas (eds.), *Investigating Science Communication in the Information Age: Implications for Public Engagement and Popular Media*. London: Oxford University Press, pp. 166–179.

Van der Sluijs, J. P., Van Est, R., & Riphagen, M. (2010). Beyond consensus: Reflections from a democratic perspective on the interaction between climate politics and science. *Current Opinion in Environmental Sustainability*, 2(5–6), 409–415.

Van Krieken, K., & Sanders, J. (2017). Framing narrative journalism as a new genre: A case study of the Netherlands. *Journalism*, 18(10), 1364–1380.

Vargo, C. J., & Guo, L. (2017). Networks, big data, and intermedia agenda setting: An analysis of traditional, partisan, and emerging online US news. *Journalism & Mass Communication Quarterly*, 94(4), 1031–1055.

Vargo, C. J., Guo, L., & Amazeen, M. A. (2018). The agenda-setting power of fake news: A big data analysis of the online media landscape from 2014 to 2016. *New Media & Society*, 20(5), 2028–2049.

Wahl-Jorgensen, K., Berry, M., Garcia-Blanco, I., Bennett, L., & Cable, J. (2017). Rethinking balance and impartiality in journalism? How the BBC attempted and failed to change the paradigm. *Journalism*, 18(7), 781–800.

Weaver, D. H., Beam, R. A., Brownlee, B. J., Voakes, P. S., & Wilhoit, G. C. (2009). *The American Journalist in the 21st Century: Us News People at the Dawn of a New Millennium*. London: Routledge.

Wormer, H. (2009). Science journalism. In W. Donsbach (ed.), *The International Encyclopedia of Communication*. doi: 10.1002/9781405186407.wbiecs016.

Chapter 7

Risk Communication

Henk Mulder and Erwin van Rijswoud

7.1 Introduction

Risk assessment is an important topic in science. Medical biologists, toxicologists, and statisticians continuously improve knowledge of risks. Chemists and engineers attempt to estimate the risks of new products and technologies as far as possible and try to control them. In thinking about the position of science and technology in relation to risk, two main roles can be discerned. On the one hand, science and technology can provide the solutions (e.g. a technology can protect someone), while on the other hand, scientific and technological developments form a potential risk in themselves (e.g. modified crops or nuclear power plants). For many people, the twofold and opposing roles of science and technology can be confusing. That is why it is so important for science communicators to understand risk and the communication about it.

Science communicators whose role includes facilitating processes which involve risk communication should first be aware of how experts, non-experts,[1] and policy-makers experience risk, or, in other words, what their perception of risk is. However, it is also important they understand the role of media, interest groups, and organizations. Armed with that knowledge, they are better

---

[1] Who these non-experts are is not always clear. In this chapter, this group includes people who do not have relevant training/education in the field, for whom risk assessment is not central to their professions, in the role of citizen or consumer. Non-experts can, however, have important experience and knowledge. They may, for example, live next to a polluting motorway, or they may have educated themselves on the topic (autodidact).

equipped to create mutual understanding and trust among those involved in the communication process. Based on a thorough understanding of the situation, they can then provide information that allows those involved to form their own opinion, or they can try to influence opinion and behavior by persuasive communication. For the latter option, professionals often rely on theories and models focused on attitudinal or behavioral change, similar to approaches used in the fields of health — and environmental communication (see Chapters 8 and 9).

In order to understand risk communication, it helps to start with what is called here, for the sake of convenience, traditional risk communication. In traditional risk communication, which is associated with the positivist and quantitative understanding of risk (Hermans, Fox & Van Asselt, 2012), the central goal is to advise policy-makers and support the public in making informed decisions about health or environmental risks. This relates to products or activities whose risks are mapped by scientific research, and the public, or part thereof, classifies the risks as either bigger or smaller than many scientists would do. One can think of the effects of chemical plants, the side effects of vaccinations, or of an unhealthy lifestyle.

Taking this traditional risk perspective, one can speak of simple risks, in which the *probability* of an event and the *effect* of this can be calculated fairly accurately by experts. However, risks also exist that are characterized by a high degree of complexity and uncertainty (Hermans, Fox & Van Asselt, 2012), such as the risk of flooding owing to climate change or the risks of new technologies, such as genetic engineering and nanotechnology. The development and application of these technologies may, for example, contribute to better diagnostics, cheaper medicines, and the production of products such as biofuels and biomaterials, but these benefits are still a promise.

Estimating how a new technology is going to develop is complex and the process is accompanied by a considerable degree of uncertainty. Promises of benefits have not yet materialized, and risks cannot be assessed in detail yet. Also, opinions differ on the ethical and social impact of these new technologies. Pidgeon & Rogers-Hayden (2007) describe this in the context of the debate on nanotechnology. The discussion of new technologies is, thus, broader than a discussion of simple risk. Chapter 4 elaborates on the way in which these new developments in science and technology, and their ethical and societal implications, e.g. complex, ill-defined wicked problems, can be deliberated and better understood through dialogue.

This chapter provides further insight into how the important aspects of risk, and the communication about it, impact decision-making on and the understanding of complex science and technology issues. It describes the various ways in which experts, citizens, policy-makers, media, and interest groups approach risks. It gives insight into risk communication processes and provides help to adequately carry out those processes. It concludes with a discussion on strategic choices to make to employ either an informative or a persuasive risk communication strategy.

7.2 Risk Considered as Probability Multiplied by Impact?

In the last few decades of the 20th century, risks associated with, for instance, radiation, chemical plants, smoking, and natural disasters, have attracted increasing attention (Covello & Sandman, 2001; Fischhoff, 1995). Researchers and scientists think quantitative calculations of risk will allow the easy transfer of findings to non-experts through a process of transmission. With the right information, non-experts can then make their own decision on how to act. However, this process does not work that way. It does not turn out to be that simple.

Often, non-experts assess the risks — especially at first — very differently, that is, more emotionally than experts. For example, when experts and consumers are asked to rank the risks of certain foods, both groups propose a different order of priority. The experts find an unbalanced diet to be most risky, while consumers think that environmental contaminants pose the biggest risk. Consumers also estimate the risks associated with additives and pesticides in food to be bigger than nutrition experts do (Peeters & Breedveld, 2009). So, a clear difference in risk perception exists.

7.2.1 The Risk Perspective of the Expert

Generally, experts express risk as *probability multiplied by impact* (considered as a negative effect). A risk is then expressed as the 'chance of dying', for example. Sometimes experts also use a concept such as 'average number of healthy life years lost'. In such cases, non-fatal injuries and illnesses can also be taken into account. Thus, risk can be expressed as a number. The calculation makes risks comparable, and limits or standards can also be set by using such figures, provided the experts have all the data needed to make the calculations. It also means that statements by experts usually deal with statistical cases and do not relate to individuals. For concerned citizens, this is less satisfactory as they primarily think of people as flesh and blood and not as statistical cases.

Full risk calculations are very complex; for example, not all data are always available or reliable. Thus, one cannot be sure whether the risk analysis covers everything. System boundaries of the analysis are always a point of discussion: which scenario should or should not be taken into account (Cohrssen & Covello, 1989)? An attack with a jumbo jet? An earthquake with a resulting tsunami? Also, human behavior is difficult to factor into calculations. If operators of a nuclear reactor do not do their job properly, the chance of something going wrong will be bigger than what the risk model predicts.

In all communications, it must be taken into account that expert knowledge may not be complete and that risks can be complex. A risk communicator must therefore be open to the fact that there may be other estimates and unknowns. All knowledge should then be discussed and assessed through a dialogue. It is important to keep in mind that some forms of knowledge, such as academic knowledge, are more easily validated than others, such as professional or experiential knowledge (see also Chapter 1). Moreover, sometimes uncertainty about risk is created intentionally by stakeholders, such as the tobacco or oil lobbies, or by activists with a broader political purpose (Boykoff & Boykoff, 2004; Gore, 2006; Oreskes & Conway, 2010; Schlichting, 2013).

Also, an Internet rumor based on a presumed causal relationship can lead to a lot of uncertainty, even after the causal relationship has been refuted by scientists. Therefore, it is important to make the difference between *correlation* and *causality* clear. A well-known example of such a rumor is the link that was suggested to exist between the MMR vaccine (measles, mumps, and rubella) and the development of autism. The perceived causality between both has long defined the public debate in Britain (Clarke, 2008), even though the publication leading to this hoax was later retracted (Rao & Andrade, 2011).

7.2.2 *The Risk Perspective of the Non-expert*

Citizens who are not experts in the field of risk and who are faced with a certain risk for the first time, usually do not react as rationally as experts. Their perception of risk is guided by a kind of gut feeling, by emotional factors called *risk perception* factors. The American psychologist Paul Slovic laid the foundations for the understanding of these risk perception factors with his research in the US in the 1970s and 1980s (Slovic, 2000). Communications specialist Peter Sandman further classified and explained these factors in his many instructional videos (Sandman, 2014). These perception factors are also relevant in many (industrialized) countries at present, notwithstanding obvious cultural differences that go beyond the scope of this chapter. The perception factors help risk communicators understand responses of non-experts to risk and even predict and possibly influence these responses.

Sandman describes risk as *hazard + outrage*. By hazard he means the risk calculated in terms of *probability × impact*, the way experts perceive risk. However, whether something is a risk in the eyes of non-experts is partly determined by a second factor, which he calls outrage. Slovic's studies basically show two main elements that explain outrage: the perceived threat (scariness) and the perceived knowledge of the risk. In addition to these components, the balance between the risks and benefits of a specific activity also plays a role in risk perception. Finally, the way in which the communication process is shaped also influences risk perception. Table 7.1 provides an overview of the perception factors that guide non-experts' initial responses to risk as either safe (more acceptable) or unsafe (less acceptable).

a. **Threat:** The perceived threat is an important element. Factors that make a risk feel threatening are, for example, the dread associated with something, such as chemical waste, or negative memories that may be evoked due to newspaper reports about a rare disease. In contrast, risks that one takes voluntarily, or risks which one has control over, are perceived as less

Table 7.1: Risk perception factors.

Perceived as safe, more acceptable	Perceived as unsafe, less acceptable
A. Threat	
Not dreaded	Dreaded
Does not evoke negative memories	Evokes negative memories
Voluntary	Involuntary
Controllable (by oneself)	Uncontrollable (by oneself)
Victims statistical	Known victims (myself, my loved ones, acquaintances, etc.).
Chronic (victims at different locations, spread over time).	Catastrophic (victims in one place and at the same time).
Few people exposed	Many people exposed
B. Knowledge	
Visible, direct effect	Invisible, delayed effect
Natural	Unnatural
Familiar (old, we are used to it)	Unfamiliar (new)
Knowledge is present (self/others)	Knowledge is not present (self/others)
C. Balance	
Substantial benefits/fair distribution of benefits and costs.	Few benefits/unfair distribution of benefits and costs.
No moral issues or risks for future generations	Moral issues or risks for future generations
D. Process	
Little media attention	Substantial media attention
No possibility of collective action	Many opportunities for collective action
Trust	Distrust
Responsive process	Non-responsive process

Note: The last four aspects are more about the process of communication than about the risk itself.
Source: Based on Slovic (2000) and Sandman (2014).

threatening. If the risk means that many people could fall victim at the same time and in the same place, people feel that the risk is more threatening (catastrophic) than if the same number of people would die over a longer period of time in various places (Birkland, 1997). As a cynical thought experiment, how would people assess the risk of smoking if all 1,300 daily US victims of smoking were to die simultaneously, every day again, at noon on Times Square, New York?

b. **Perceived knowledge:** The heading of perceived knowledge contains more perception factors than just the explicit factor of knowledge itself. Someone who does not know exactly what the risk entails feels that this risk is bigger. This effect also occurs if knowledge seems to be absent in those responsible for controlling the risk. A number of other factors contribute to concerns about the effect of a lack of knowledge on a risk. People perceive invisible effects, such as radiation and the cancer it may cause, as riskier because they deal with uncertainty less well than with bad news. If the risk has been around for a long time, people become used to it. That is why risks at home are underestimated. It is at home where many accidents take place: accidents that occur through carelessness while using a stepladder, for example. For the same reason, the risk of a natural disaster is estimated as lower than it is, although here a moral issue can also come into play — the idea that mankind has to accept natural (or God-given) risks. Also, people have a more positive feeling about natural products than about artificial, industrial, or unnatural ones. However, as stated before, of course many cultural differences in perception can also apply here.

c. **Balancing:** When considering the acceptability of a risk, people weigh up the advantages (benefits) and disadvantages of the activity that causes the risk and decide whether these are fairly distributed. As a result, personal benefits often tend to carry larger weight than collective or social benefits. This means that health (personal) and environmental (collective) risks, for example, are approached differently.

d. **Process:** In addition to factors associated with the risk itself, elements of the communication process also play a role in the perception of the risk. The outrage about a risk often increases if activist groups communicate about it or if it is in the news a lot. Alternatively, people experience a lower risk if they have confidence (and trust) in those who manage the risk and communicate about it, and if the communication process itself offers them a meaningful way to influence the management of risk. This is called a responsive process, which is discussed later in this chapter (Section 7.5).

On the basis of insights into these perception factors, risk communicators can now understand the difference in the ranking of the risks associated with food, as was discussed in the introduction to this chapter. They can also understand other risk discussions[2] and answer the following questions: smoking is a high-risk activity according to experts, but why do smokers assess this risk differently? How does the discussion proceed when second-hand smoke is included? Taking into account the perception factors, it is possible to explain why many countries have a smoking ban in public areas and in bars and restaurants. The discussions that have occurred in the Netherlands about whether to allow smoking in small pubs without employees other than the owner can also be understood.[3] Finally, it is also possible to explain the effect of messages which would indicate that the tobacco industry adds addictive substances to tobacco.[4]

Understanding the perception factors assists the process of risk communication by which non-experts can be helped or influenced. One simple example comes from a Dutch student of science communication, who interviewed someone about vaccinations. The interview illustrated that the interviewee had assumed that artificially made antibodies were injected. When the student explained that in fact the injection was a weakened natural virus and that the body makes the antibodies itself, the interviewee's perception of the risk changed and she also perceived vaccination to be less risky. The risk was now natural rather than unnatural. That the message could be communicated might also be because trust had developed between them and the student was seen as a neutral — very trustworthy — source. After all, the government had been sending out similar messages, which opponents merely undermined by putting the naturalness of vaccination in doubt. Although information was not the purpose of the interview — let alone persuasion — the student arrived in the classroom the following day quite happy. She had done her first risk communication! An understanding of the *mental model*

---

[2] However, no general ranking system exists to indicate which perception factor is the strongest. This varies from case to case and many case studies are available in the literature. For an overview, see, e.g., Sjöberg, Moen & Rundmo (2004).

[3] See: https://www.centreforpublicimpact.org/case-study/smoking-ban-netherlands-2008-amendment-tobacco-act/

[4] See e.g. https://www.jellinek.nl/vraag-antwoord/worden-aan-tabak-stoffen-toegevoegd-om-de-kans-op-verslaving-te-vergroten/ (last visited 11/22/2019)

that another person has of the risk is, thus, very important for effective risk communication (Morgan *et al.*, 2002).

7.2.3 *When Non-experts Assess Risks as Being Bigger Than Experts*

People assess some risks as being bigger than experts do, for example because something is new (see Table 7.1). A comparison with an old and well-known risk can then offer them an anchor. A risk communicator should, however, not make the mistake of comparing incomparable risks to each other. For example, although an incinerator poses less danger to residents than smoking, the latter is voluntary, whereas an incinerator in the neighborhood is enforced. Making such a comparison could, thus, lead to outrage.

Much has been written about making risk comparisons and using them as a tool in risk communication. Sandman (2014) provides numerous guidelines for good comparisons. But most important in risk communication is that communicators must ask themselves whether they really want to help someone or whether they want to steer people in a certain direction or even want to ridicule their fear. It helps if communicators are perceived as independent sources, such as the student who had done her first risk communication. An independent source may often be permitted to compare different types of risk; for example, in cases where a representative of an industry making the same comparison would not be shown the same courtesy. Any source with an interest will be considered less reliable. In such cases, openness is the best policy: 'I acknowledge that I have an interest here, but nevertheless...'.

Communicators will need to acknowledge people's outrage about the risks to which they are exposed before they can engage in conversation. If they fail to do so, they may turn against them. Empathy and open communication also help. Communicators can show their understanding explicitly, but they can also express it through their posture and the use of short phrases that indicate they do understand. They should acknowledge that people may be worried and are allowed to feel this way (e.g. outrage because something is involuntary), however minimal the exposure may be. Please note that to be sincere and trustworthy is very important. Only then communicators can gain the confidence and trust that allows them to explain the views of experts on *probability × impact*.

Non-experts have proven themselves perfectly capable of engaging themselves in the assessment of a hazard when it is of great importance to them. Then they rely less on their gut feelings. Therefore, it is certainly not a question of incapability. Good examples are the decisions that patients must make about the risks of an illness vs their treatment (Lorig *et al.*, 2001), how patients contribute to the development of knowledge in the medical field (Epstyein, 1995), and the speed at which neighborhood groups can become acquainted with many facets of risk (own experience of author, H. Mulder). A classic study by Brian Wynne described the value of farmers' local knowledge in dealing with fallout after the nuclear disaster in Chernobyl (Wynne, 1992).

In risk communication, an expert will always need to address the possible impact of a problem, even if the probability of a specific impact is small. For example, one of the authors of this chapter attended an information evening on chlorine transport by train, through the city. The message of the public authorities was: 'according to experts, the chance that something will go wrong is very small. There will only be an incident once every 10,000 years, which is much less than in the United States, because our trains have different coupling systems. So what are we worrying about?' The people who lived near the tracks felt very upset and were outraged because they felt that they were not being taken seriously. Also, the officials did not offer an emergency action perspective (unlike recommendations in, e.g., Covello & Sandman, 2001). They could not answer questions such as the following: 'Should we evacuate or stay inside?', 'Should we go to the attic or the basement?' 'Should we stand in the shower?' This reinforced the perception of a lack of control and confidence/trust, causing the outrage to increase.

Providing people with an action perspective gives them more 'control' over the situation. They gain more knowledge, which leads to a reduction in their sense of outrage. It can also prevent stress-related symptoms.[5] Therefore, provision of an action perspective can be seen as valuable in all risk communications (Gezondheidsraad, 2014).

A good and very simple example of better communication through offering an action perspective can be seen in the changes to communication

---

[5] Stress, with all its associated physical symptoms, can occur if one cannot escape or control a situation. In this case, escape by moving is not a real option, and if one does not know how to manage an emergency or to escape from it, stress may occur.

about fires in the Netherlands. Until a few years ago, the message was often: 'there is no danger to public health'. This is a vague message which concerns overall health in one area and not an individual's health. Today, fire engines bear the message: 'stay away from the smoke.'

7.2.4 *When Non-experts Assess Risks as Being Lower Than Experts*

Some risks are considered by non-experts as low, whereas according to experts, there is great danger. Examples are eating fatty food, which is very normal and natural, or driving under the influence, during which people still think they are in control (see Table 7.1). When communicating in this case, communicators may aim at perception factors that could increase outrage. Thus, in campaign materials such as videos, victims may be portrayed as people one can relate to, as victims that could be one's child, for example. However, simply creating fear does not work, communicators must also offer people an action perspective (Kok *et al.*, 2013). How can people avoid the risk? This type of approach can be seen in traditional, public mass media campaigns on healthy eating, daily exercise, and the campaigns for designated drivers. A similar approach is used in a Mexican travelling exhibition where local people underestimate the dangers of the ashes of local active volcanos (see Box 7.1).

How far communicators choose to go in influencing people is an ethical consideration. Where does free advice stop and imposed choice start? This question will be dealt with in-depth at the end of this chapter (see also Chapter 8). When communicators use persuasive communication, they challenge someone's autonomy. If authorities initiate persuasive communication, this only seems justified for important issues, where no alternative to persuasive communication exists. Van Woerkom (1988) gave some easy-to-use guidelines for this, proposing that persuasive communication is permitted in situations where detection and punishment are difficult, such as cases where people are flushing used cooking oil or motor oil down the toilet, or disposing batteries with household waste. According to him, it may also be that another policy is too expensive: for example, it is better that people take their waste with them from the forest because the municipality cannot place bins everywhere. Finally, other measures may take too much time; for example, when a disaster occurs, people in the affected area should immediately go

Box 7.1: Communicating about the unknown dangers of volcanoes

Text: Dr Elaine Reynoso, researcher/science communicator, National Autonomous University of Mexico (UNAM), Mexico.

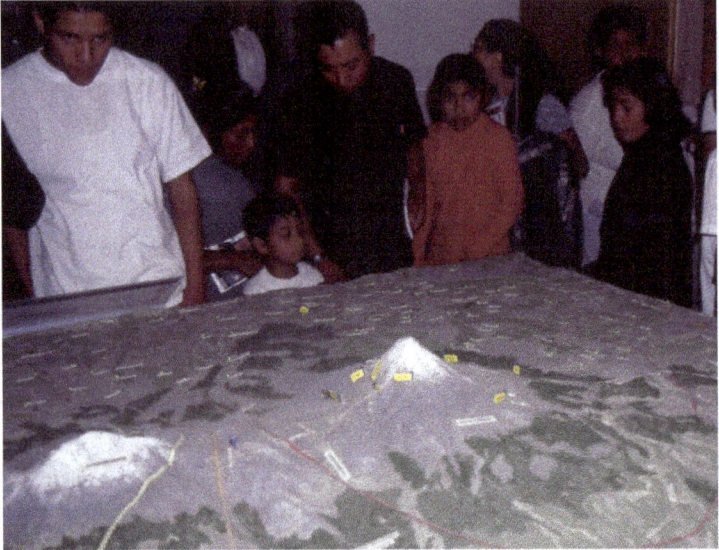

Photo credit: Arturo Orta.

After a heavy rainfall of dangerous volcanic ashes of the Popocatépetl on Mexico City in 1997, a travelling exhibition was made in order to inform the people living in Central Mexico about the risks of an active volcano in their living environment. In the development of the exhibition, differences in risk perception were taken into account, adequately addressed, and an action perspective was offered.

Popocatépetl means smoky mountain in Náhuatl (the local indigenous language). This volcano and its companion *Iztaccíhuatl* (which means *'white lady'* due to the snow on its peek) are an outstanding feature of the landscape of central Mexico in a densely populated area, which includes several large cities such as Mexico City, Puebla, Cuernavaca, and many towns. The volcanoes are part of the local culture and have been an inspiration for novelists, painters, photographers, and filmmakers. They are also very popular with mountain climbers. It is difficult to imagine that one of these beloved volcanoes can be so threatening.

Box 7.1: (*Continued*)

Volcanic ashes when dry have a light powdery consistency and appear to be quite harmless. However, they are extremely dangerous. These ashes are very abrasive; can cause great harm to eyes, respiratory system, and electrical appliances; and are a threat to traffic and airplanes. When these ashes get wet, they form a very dense and heavy substance which can collapse roofs and cause a severe blockage of the sewage system if they are swept down the drain.

Although after the heavy rainfall of ashes, constant information was provided through different media advising people on what to do with the ashes, the fact that the common word *ashes* was used, made it difficult for people to understand the potential danger. Therefore, the National Autonomous University of Mexico (UNAM) created a small travelling exhibition with the purpose of providing Mexico City residents with the necessary explanations and advice on what to do before, during, and after a rainfall of volcanic ashes. This action perspective included aspects such as acquiring protective gear, covering water tanks, trying to sweep everything up and collecting it in bags, dusting the ashes off the surfaces when dry, and not throwing the ashes down the drain.

Front-end evaluation for planning this exhibition included an analysis of how the media were dealing with the issue and interviews with inhabitants in this region, with the purpose of inquiring about their knowledge on the topic, their fears, interests, alternative frameworks, and understanding of terms such as volcanic ashes.

In the year 2000, Popocatépetl had a major eruption. Other cities and towns in the risk region wanted the exhibition. Therefore, it was expanded to include other volcanic risks as well as the new knowledge acquired through years of monitoring (Reynoso, 2003).

The following were the messages of this new exhibition: volcanic activity is a consequence of our ever-changing planet, disasters are caused by the lack of knowledge and prevention, and science provides the necessary knowledge for action. The exhibition provided information about volcanoes and their activity, how the volcano is monitored, and instructions on what to do in the case of an eruption. Last, but not least, it had an emotional approach with the history, stories, legends, culture, and art related to the Popocatépetl and its eternal companion Iztaccíhuatl.

inside. Likewise, persuasive communication and compulsory education are still the best measures for AIDS prevention, as a cure is not yet available.

Research also shows that communication works well in conjunction with accompanying policies; for example, an alcohol and smoking campaign combined with a policy designed to discourage it through taxes, age limits, and good role models (Van Woerkom, 1988; see also Chapter 8). Finally, the message must also fit within one's social orientation (Kahan, 2012). The more it differs from the current views held by an individual, the more difficult it will be for them to accept the message. Communicators should also be aware that people do not actively seek information that refutes their existing ideas. Rather, it is the other way round. People are open to information that confirms their own ideas. This is called 'confirmation bias'[6] (see also Chapter 10).

For a good communication strategy in such cases, communicators must also understand the factors that govern human behavior; in other words, it is important to have some knowledge of behavioral theories (see also Chapters 8 and 9).

7.3 Risk Perspective of the Policy-maker

Policy-makers (and company directors) try to reduce risks as much as possible, but they do not have an unlimited budget. A general aim in risk policies is to limit fatalities per year, e.g., to the often used one in a million (mortality 10^{-6}) as a result of a particular activity or product. Thus, a problem arises when a policy-maker is forced to take measures against high-tension wires or mobile phone masts near homes, due to public pressure, because they are felt to be 'unsafe' (see Table 7.1).

As an example, the Dutch National Institute for Public Health and Environment (RIVM, 2003), argued that high-tension wires cause at most one additional death per 15 million per year in the Netherlands, based on an unproven potential relationship with childhood leukemia. Technical adjustments to the current electricity transmission system, or moving pylons or homes, could easily lead to an expenditure of between EUR 3.5 million and

---

[6] A term coined by Wason in 1960 (Silliman & Wear, 2018).

EUR 1 billion per death avoided (USD 3.92 million and USD 1.12 billion).[7] In contrast, in relation to traffic accidents, one can avoid one death for a cost of EUR 1,000 to EUR 1 million (USD 1,120–1.12 million). However, people are less concerned about traffic risk, which is an old or familiar risk, with 75% of people also thinking that they are a better than average driver[8] and therefore have control (see Table 7.1).

In assessments of medical interventions, similar discussions are present about the acceptable costs of gaining one healthy life year (which in the US is generally taken to be in the range of 50,000 dollars, though in practice higher amounts are sometimes invested or lower amounts not funded (see e.g. Neumann, Cohen & Weinstein, 2014)).

7.4 Role of the Media and Interest Groups

Non-experts' initial reactions to risk often come about intuitively and, thus, they may easily find support and gain a considerable response. This effect is called the *social amplification* of risk. Nowadays, confusion and restlessness can spread even faster through social media (Gezondheidsraad, 2014). A small incident that is new or frightening then creates so much social upheaval that the stakeholders may even suffer a financial loss or an unnecessarily strong sense of insecurity in society may result. For example, a report on the potential damage from UMTS/4G mobile phone masts may have financial consequences for providers or a few cases of infectious disease may lead to a considerable disruption of passenger traffic.

In addition to amplification, a weakening of a risk message can also take place (*attenuation of risk*). This process seems more passive. In this case, the message concerning a risk which is felt to be safer (based on the perception factors) is simply not passed on. Attenuation can also occur when active stakeholders find a certain message unwelcome (e.g. tobacco lobbyists who

---

[7] This example becomes more complex when one considers that experts disagree (a) on whether the risk exists at all and (b) on whether one should divide the number of deaths by the whole population of the Netherlands or only by the number of people in the 23,000 properties close to the power lines. This issue is beyond the scope of this book.

[8] See, e.g., McCormick, Walkey & Green (1986) for this effect, which is an example of a psychological mechanism called 'illusory superiority'.

downplay the dangers of smoking) or genuinely disagree with a message about the high risk. For more about the *Social Amplification and Attenuation of Risks Framework*, see, e.g., Hermans, Fox & van Asselt (2012); Kasperson & Kasperson (1996); Pidgeon, Kasperson & Slovic (2003); Renn (1991, 2011).

Although the German risk assessor Renn (1991) has shown that social responses are often carefully considered and that panic seldom ensues, the notion of social amplification does provide an insight into how broader communication processes related to new risks develop. It seems that in such cases relatively minor incidents involving some applications of new technology are seen as harbingers of possible catastrophes. Activist groups may also join the discussion and increase the perception of risk. Societal organizations often make more use of personal stories and, thus, their message can strike harder than that of governments or experts who express risk in terms of abstract probabilities. Advocacy groups do, however, vary widely, and some may engage more readily in a substantial scientific debate than others. For the risk communicator, it is important to maintain good communication lines with stakeholders and attempt to gauge in advance how the message will come across.

The role of the media in the communication process is very important (Boykoff & Boykoff, 2004; Clarke, 2008). Only the unusual is news and, thus, stories about new or very rarely occurring hazards often make the headlines. This is also one of the reasons why people remember precisely those kinds of risk and perceive them as bigger. These are the negative memories that are evoked (see Table 7.1). In his study of perception factors, Slovic (2000) presented respondents with a list of different risks and asked them to indicate which risk claimed most victims. The respondents overestimated the mortality for the least likely risks, such as botulism and vaccinations. These are so rare that they are reported in the newspaper. In contrast, respondents underestimated bigger risks, such as the chance of getting diabetes.

In addition, a bias in favor of negative news, such as accidents and death, exists. The real stories behind the headlines on the front page are often only found elsewhere in the paper. Moreover, the media are often most interested in conflicts between stakeholders (Semetko & Valkenburg, 2000; see also Chapter 6).

All of this can trigger a process of social amplification, where activist groups create a conflict, the press writes about it and the stakeholders are, thus, forced to respond. However, the media should, as a journalistic ethical requirement, also listen to both sides of the story. Attention is, thus, paid to various opinions about the risk. It may then appear as if the knowledge available is low because experts apparently disagree, which may lead to the feeling that the risk cannot be controlled. However, the opposite can also occur. Placing equal emphasis on both sides of the discussion can lead to the conclusion that the risk is not that high or urgent, for example, when the so-called skeptics receive a disproportionate amount of attention in the media reports on climate change (see also Chapter 6). The media may, thus, amplify or attenuate risk.

Social media can reinforce and speed up this social amplification. Shocking stories are quickly shared. YouTube videos of activists talking about the dangers of vaccinations are easily found. Another example is the pandemic influenza in 2009, which was framed as a conspiracy of pharmaceutical companies to sell more vaccines, thereby undermining trust in health authorities (World Health Organization, WHO, 2013). Similar panic induction was seen in a number of countries with the introduction of the human papillomavirus (HPV) vaccination for girls aged 12, aimed at preventing cervical cancer. However, later on, the opposite occurred in the Netherlands, demonstrating that rumors can also be unmasked quickly through social media. The girls themselves appeared to be especially afraid of the length of the needle for the HPV vaccine and the pain it would inflict. During the later HPV vaccination campaigns, girls put each other at ease through social media. The group that had been vaccinated explained to their peers that it was not so bad (Timmer, 2010). This is an example of the social attenuation of a risk.

Social media can also play a role in making real threats more rapidly known, as was the case for the H7N9 virus outbreak, which was identified on Weibo, the Chinese Twitter (WHO, 2013). The Dutch Health Council adds to this that in general, through more openly available scientific information and through citizen science projects (see also Chapter 5), citizens themselves can produce and share data that can conflict with official data (Gezondheidsraad, 2014).

The exact influence of modern social media on risk communication has not yet been studied extensively. Neely (2014) concludes that risk communicators should embrace social media, as they make the difference between telling people facts and being heard. Similarly, the WHO calls on health authorities to be better prepared and active on social media (WHO, 2013).

7.5 Responsive Process

By being open and by responding to the questions and fears of those involved and by giving stakeholders some influence on the decision-making process, communicators can establish risk communication as a *responsive process*. In a responsive process, participants listen to each other and act on the outcomes of the consultation. Various studies show that responsive, participatory processes in, e.g., environmental decision-making create more legitimate, higher quality decisions (see also Box 7.2), provided a well-performed, genuine, timely, and sustained engagement is present (National Research Council, 2008).

Box 7.2: A responsive process in practice — community support for liquefied natural gas (LNG)

Text: Barbara Campany, Senior Technical Director, GDH, Australia.

This example of a responsive risk communication process comes from my work experience as an environmental risk communication practitioner with Stakeholder Engagement and Social Sustainability at GHD. GHD is a professional service provider in the global market sectors of water, energy and resources, environment, property and buildings, and transportation in Australia. My work often focuses on projects requiring legislative approval for development.

One of these projects included gaining community support for the approval and construction of a LNG storage facility in a regional community in northern New South Wales, in Australia between 2011 and 2013 (the installation and risk is non-natural, involuntary; and issues relating to controllability, unfamiliarity, knowledge, distribution of benefits may be present, thus potential outrage could result). Apart from the standard environmental concerns these types of projects trigger, one of the most emotive issues was the potential loss of koala habitat as part of the land clearing for the LNG facility. Koalas are a loved and endangered species in Australia and special interest groups passionately campaign for their protection (so the potential

---

Box 7.2: (*Continued*)

victims, even though not human, are recognizable, not statistical). Using Peter Sandman's risk communication approach helped guide the timing for conversations with the koala (and other fauna) interest groups.

The company was committed to providing a process that gave the locally affected community and interest groups scope to meaningfully engage, so that issues raised by them would be considered early in the environmental assessment. Another of the company's commitments was to provide an inclusive process that demonstrated involvement and disclosure that could nurture a trustworthy relationship between the company and the host community (including local interest groups) over time.

Those locals who cared most about the potential impacts were typically outspoken and somewhat guarded about the company's commitment to frank, open, and inclusive engagement. Ongoing stakeholder participation in several workshop-style meetings helped identify key environmental concerns and jointly determine management and mitigation processes that allowed the relationships to prosper (trust development). The commitment was to ensure that concerns were identified and management processes were jointly agreed upon and then articulated in the conditions of assessment. This proved effective as community and interest group inputs were considered as part of the process, rather than the company completing an assessment in isolation from the community and interest groups, and then presenting it as a *fait accompli* for community consultation. In effect, the community and interest groups participated in establishing some of the assessment criteria.

Significant contribution by these stakeholders through numerous briefings strengthened the flora and fauna management plans regarding issues where their main fears manifested, for example, making sure fauna had enough time to relocate nests with minimal stress prior to the site-clearing process, undertaking scatology studies to confirm presence of koalas, and for the company to provide other lands dedicated to koala habitat into perpetuity. Because the stakeholders had significant influence over the environmental conditions being applied, the risk of outrage diminished.

These approaches provided shared ownership of the outcomes. They gave credibility to the process and accountability to the company. These are all features of Sandman's Outrage Management model of risk communication.

When organizing a responsive process, it is important that it has a real potential to influence the policy on and perception of risk (Sandman, 2014). It is also of great advantage to be aware of the history of contact between those responsible for the risk — or the communication about it — and the various stakeholder groups. Sometimes, the source has a bad name and a communicator, who is the representative of the source, cannot simply start with a clean slate. It helps when non-experts can be empowered by having access to independent expert backup, e.g., through Science Shops (Mulder & De Bok, 2006), especially when trust levels are low (Kasperson, 2014).

One issue present in all discussions about risk is that sometimes it is not clear who is responsible for the potential problems. Some government agencies are not able to monitor adequately whether companies comply with licensing requirements. It may be unclear who exactly is responsible for what. With whom should one talk and who controls the risk? (cf. *organized irresponsibility*, Beck, 1994). Clarifying matters is an urgent task for the government and, thus, part of risk communication (Gezondheidsraad, 2014).

In a responsive process, a joint assessment can determine which risks are acceptable. Citizens do not necessarily aim for zero risk. On the basis of ethical considerations, a particular risk in a given situation may be quite acceptable. In such cases, the discussion will concern the fair distribution of benefits and risks rather than whether the risk is one in a million or a little lower or higher. This has important implications for communication. The framework shifts from the claim that 'the risk is very low', with a focus on convincing people of this, to 'this risk is acceptable because there is a just, fair distribution of burdens and benefits, e.g. by compensation' (Van Eeten *et al.*, 2012).

Setting up a responsive process is also fundamental to handling more complex risks, such as climate change, or those risks associated with new technologies, such as nanotechnology and synthetic biology (see also Chapter 4). Communicators facing the challenge of communicating about complex risks in practice can use all the help they can get (see Box 7.3 for a few pointers).

Box 7.3: A few pointers for risk communication practice

Technical Information: Too much technical information can lead to people not seeing the wood for the trees. So, simplify, but sometimes providing full information is obligatory (legally). Therefore, it may work best to provide the information in a layered way: abstract, main text, appendices. On a website, one can work using terms such as 'in brief', 'read more', and 'references/links', thus moving toward complete information.

Jargon: Pay special attention to technical terms that have a different meaning in everyday language. For example, 'exposure' is a word that is associated by the public with direct damage, but for toxicologists it is not. 'Significant' is a statistical concept, but if a communicator says that cancer is not significantly present in a neighborhood, it may lead to outrage. If in the Netherlands there is a 'moderate smog' alert, it means it is very bad and Cara patients should be aware. Also, in the Netherlands, 'serious' soil pollution does not always mean that urgent treatment is required. Sometimes, the pollution is situated under a layer of concrete and is contained and therefore not dangerous.

Sources: Based on Sandman (2014); Gutteling & Wiegman (1996); personal experience of author H. Mulder.

A lot has been written about the effects of framing and the use of narratives (story-telling) and emotion in messages and communication processes. These concepts are of broader relevance to science communication and discussions and go beyond the scope of this chapter (see, e.g., National Academies of Sciences, Engineering, and Medicine, 2017; Roeser & Fahlquist, 2014; Chapter 6).

7.6 Directive or Responsive Communication?

In the approaches described above, risk communication seems difficult and complex, although not entirely impossible. But, how to approach the communication process? One route involves influencing behavior through persuasive communication (see also Chapters 8 and 9). Another route involves giving citizens the freedom to decide for themselves and to talk about risk through a responsive process.

Risks are evaluated differently by experts and citizens. It is important to have insight into how citizens understand the risks before starting the communication process. However, the process does not always lead to an awareness campaign. It is valuable to know that multiple interpretations of the risks are possible, especially when broader ethical, legal, and social aspects play a role. In such cases, setting up a social dialogue may be more sensible than starting an awareness campaign.

By taking steps toward acknowledging the ethical and social aspects, risk communication may be viewed from a broader perspective. This approach recognizes that risks are uncertain and complex and cannot be resolved by a single solution. In addition, clashes of interest and values between the various stakeholders involved may be present, and consequently, one clear action perspective cannot be presented. To illustrate these dilemmas, an example is given below, where a *directive approach* and a *responsive approach* are introduced as two possible extremes of the communication process.

Two approaches to communication: *Imagine*: Recently, a dangerous variant of bird flu has surfaced in the region and experts say it is contagious to humans. The government decides to vaccinate. A communications expert at the National Immunization Program (NIP) has been commissioned to design a communication campaign for this vaccine, which will be offered to children, as well as others, and involves a well-tested vaccine. The active substance has been used for a long time, but a modified variant will be used to combat this particular flu virus. The agents added to increase the effectiveness of the vaccine are new and are being used for the first time on humans. What would a directive and a responsive communication strategy look like?

Approach 1: Directive strategy

In this first scenario, the NIP aims to obtain the highest possible vaccination rate via the reduction of false perceptions of risk that might deter people from vaccination. That new agents are used is of secondary importance. Research has already shown no risks to the population.

In addition to informing the audience about safety, how people think about the vaccine and how they estimate and experience the risk are actively monitored. This knowledge can be used to achieve the goal, thus lowering the outrage over the vaccination risk. The NIP believes in the education of the public in order to remove misconceptions about vaccination in the public domain. Precisely because information and sources of information can be complex, misleading, or inaccurate, citizens must be protected against this. It is not just about having a good understanding of the government's campaign but also about filtering information that reaches the people through many other channels, such as the Internet, newspapers, and television, which can present risks incorrectly. This is even more necessary when action groups construct fables about side effects and mobilize a counter-campaign, thereby undermining the scientific authorities.

In this scenario, therefore, the emphasis is on citizens' inability to separate the pros from the cons. In addition, both existing and new means of information provision and guidance are actively deployed to convince citizens. The success of the communication campaign around the national vaccination program can then be directly inferred from the number of people within the target group who have the vaccination (the degree of vaccination).

The role distribution is thus: 'the public health care system diagnoses the problem of poor reception of its health advice primarily as a technical problem — "interventions do not work" — and seeks solutions primarily in "more knowledge about risky behaviour", so that citizens can be directed toward more healthy behaviour' (Horstman, 2011).

Approach 2: Responsive strategy

In the second scenario, the NIP chooses a different route. The choice not to vaccinate is not seen as an expression of an incorrect or incomplete understanding of government information, but as the result of a legitimate personal view, which can be related to both the supply of the vaccine itself and the information about the vaccine. Respecting the concerns of the target group and their choice to decline is important here and it is just as important to recognize that existing scientific knowledge also leaves many questions unanswered. Not all risks are equally well documented. Furthermore, the concerns and criticisms of citizens are not only considered at the level of information or risk perception but also in terms of interaction. The focus lies on the way public concerns find a place and may be shaped in the interaction with experts.

In this scenario, good decision-making by parents is something that the NIP can contribute to, by examining the criticisms and concerns of citizens with an open mind and communicating these to scientists. The scientists can then respond to them and this creates a responsive process. A successful vaccination program then manages to bring the parties closer together, regardless of the outcome.

The frame used by the communicator is now: Everybody may make their own decision and we would like to help them with their choice. Based on current, up-to-date knowledge, the NIP is confident that this vaccine is safe and will prevent a lot of suffering, but we are open to questions and concerns and we acknowledge that there may be other considerations that influence your decision.

The question evoked by both scenarios is: When to apply which approach? The stories indicate that it is not just a question of how to communicate but also why a communicator would choose a directive or a responsive approach? Among the many considerations that might apply, two important ones are as follows: the role of professional identity of the communicator and the question of who actually determines how the risks are framed.

In examining the role of a communicator's professional identity, the question arises as to whether the risk perception of the target audience may be different from the communicator's. What space do citizens have, to make a choice which differs from the one preferred by the communicator (or the client)? In the first scenario, communication and interaction are rather directive and the space to make a different choice is limited: 'Urgent health interests are at stake and if the situation is urgent then people must do something immediately, without discussion'.

In the second scenario, much more space is available. Simultaneously, the communicator might think that in the second scenario, the effectiveness of disease control is reduced. This does not fit with the aims and nature of the organization for which the communicator works. Preventing many illnesses and potential deaths from flu has the highest priority for the organization, and the first communication scenario is most appropriate for that, or so it seems.

Why then would the communicator still choose the second scenario? Aside from the question of the urgency of the vaccination program, there is also the question of who determines that the flu is especially threatening. In addition, the communicator may wonder whether no better ways to deal with the disease exist and whether the new vaccine is safe. Is the scientific knowledge complete? How does one weigh up the pros and cons? Is that for the communicator to decide, based on intuition or experience? Or will the communicator leave this decision to scientists or employers? Or maybe to the people who are invited for vaccination? Will the communicator allow them to make their own choices? When controversies arise in the media or in the online world, it is more difficult to direct people's attitudes. From this perspective, the second scenario no longer seems so strange because, with a responsive communication plan, the communicator can anticipate discussions. Moreover, if it is clear that the vaccine may be taken voluntarily, the outrage decreases and the perception of the risk is lowered. To quote Sandman (2012, p. 27): 'The right to say "no" makes saying 'maybe' a lot easier'. In both scenarios, trust still plays an important role. Trust in medical doctors is still high: when they recommend a vaccination, many people still consider this sufficient reason to act.

7.7 Conclusion

This chapter shows that it is very important for the risk communicator to know how stakeholders think and feel about the risks. An immersion into the viewpoints of the different groups involved is necessary. What do they know, or think they know? What do they feel? What do they want to achieve? Which risk perception factors play a role? Can an action perspective be provided? How much directive to provide? With the answers to these questions in hand, a communication strategy can be set up.

As elaborated in this chapter, how experts, citizens, policy-makers, and the media deal with risks may be very different. Experts conceive of risk in terms of *probability* × *impact*. The first reaction of citizens is related to perception factors based on the context of the risk: how threatening or unknown the risk is, whether it is fair, and how much trust there is in the source (of the risk or the communication). The expert and non-expert perspectives on risk are often opposed. What one estimates as a high risk, the other estimates as a lower risk. Policy-makers are faced with these different perspectives on risk. They have to construct their risk policy within financial constraints, which determines the communicative challenge. Media and interest groups may amplify the sense of something being risky, but they can sometimes attenuate the perceived risk as well.

Risk communicators have to work in between these often-opposed perceptions, but they may be able to facilitate a responsive process based on their knowledge of, and respect toward, the other stakeholders' positions. However, before the risk communication process starts, a decision has to be made on whether a directive approach in communication seems appropriate, or a responsive approach, in which there is more space to take the perspectives of citizens into account. Of course, a large gray area in-between exists, and good reasons for both extremes may exist. Remember that choosing a perspective that clashes with the interests of the target audience can lead to the realization of the opposite of what is intended.

As this chapter is necessarily limited in scope, the book *Effective Risk Communication*, by Árvai & Rivers (2014), is recommended for further reading.

References

Árvai, J., & Rivers, L. (eds.), (2014). *Effective Risk Communication*. London: Earthscan.

Beck, U. (1994). *Ecological Politics in an Age of Risk*. Cambridge: Polity.

Birkland, T. (1997). *After Disaster: Agenda Setting, Public Policy, and Focusing Events*. Washington D.C.: Georgetown University Press.

Boykoff, M. T., & Boykoff, J. M. (2004). Balance as bias: Global warming and the US prestige press. *Global Environmental Change, 14*, 125–136.

Clarke, C. E. (2008). A question of balance: The autism-vaccine controversy in the British and American elite press. *Science Communication, 30*, 77–107.

Cohrssen, J. J., & Covello, V. T. (1989). *Risk Analysis. A Guide to Principles and Methods for Analyzing Health and Environmental Risks.* Washington, DC: US Council on Environmental Quality, Executive Office of the President.

Covello, V., & Sandman, P. (2001). Risk communication: Evolution and revolution. In A. Wolbarst (ed.), *Solutions to an Environment in Peril.* Baltimore, MD: John Hopkins University Press, pp. 164–178.

Epstyein, S. (1995). The construction of lay expertise: Aids activism and the forging of credibility in the reform of clinical trials. *Science, Technology, & Human Values, 20,* 408–437.

Fischhoff, B. (1995). Risk perception and communication unplugged. Twenty years of process. *Risk Analysis, 15,* 137–145.

Gezondheidsraad (2014). *Risicocommunicatie op een nieuwe leest schoeien.* Den Haag: Gezondheidsraad.

Gore, A. (2006). *An Inconvenient Truth.* Hollywood: Universal Pictures.

Gutteling, J., & Wiegman, O. (1996). *Exploring Risk Communication.* Dordrecht: Kluwer.

Hermans, M. A., Fox, T., & Van Asselt, B. A. (2012). Risk governance. In S. Roeser, R. Hillerbrand, P. Sandin, & M. Peterson (eds.), *Handbook of Risk Theory.* Berlin: Springer, pp. 1093–1117.

Horstman, K. (2011). Leren van rokers en dikkerds. *Tijdschrift Sociale Geneeskunde, 89,* 139–141.

Kahan, D. (2012). Cultural cognition as a conception of the cultural theory of risk. In S. Roeser, R. Hillerbrand, P. Sandin, & M. Peterson (eds.), *Handbook of Risk Theory Epistemology, Decision Theory, Ethics, and Social Implications of Risk.* Dordrecht: Springer.

Kasperson, R. (2014). Four questions for risk communication. *Journal of Risk Research, 17,* 1233–1239.

Kasperson, R. E., & Kasperson, J. X. (1996). The social amplification and attenuation of risk. *Annals of the American Academy of Political and Social Science, 545,* 95–105.

Kok, G., Bartholomew, L. K., Parcel, G. S., Gottlieb, N. H., & Fernández, M. E. (2013). Finding theory- and evidence-based alternatives to fear appeals: Intervention mapping. *International Journal of Psychology, 49,* 98–107.

Lorig, K. R., Ritter, P., Stewart, A. L., Sobel, D. S., Brown, B. W., Jr., Bandura, A., Gonzalez, V. M., Laurent, D. D., & Holman, H. R. (2001). Chronic disease self-management program: 2-year health status and health care utilization outcomes. *Medical Care, 39,* 1217–1223.

McCormick, I. A., Walkey, F. H., & Green, D. E. (1986). Comparative perceptions of driver ability — A confirmation and expansion. *Accident Analysis & Prevention, 18,* 205–208.

Morgan, G. M., Fischhoff, B., Bostrom, A., & Atman, C. J. (2002). *Risk Communication. A Mental Models Approach.* Cambridge: Cambridge University Press.

Mulder, H. A. J., & De Bok, C. F. M. (2006). Science shops as university–community interfaces: An interactive approach in science communication. In D. Cheng, J. Metcalfe, & B. Schiele (eds.), *At the Human Scale — International Practices in Science Communication.* Beijing: Science Press, pp. 285–304.

National Academies of Sciences, Engineering, and Medicine (2017). *Communicating Science Effectively: A Research Agenda.* National Academies of Sciences: Washington, DC, The National Academies Press. doi: 10.17226/23674.

National Research Council (2008). *Public Participation in Environmental Assessment and Decision Making.* Washington, DC, The National Academies Press.

Neely, L. (2014). Risk communication in social media. In J. Árvai & L. Rivers (eds.), *Effective Risk Communication.* London: Earthscan, pp. 143–164.

Neumann, P. J., Cohen, J. T., & Weinstein, M. C. (2014). Updating cost-effectiveness — The curious resilience of the $50,000-per-qaly threshold. *New England Journal of Medicine, 371,* 796–797.

Oreskes, N., & Conway, E. M. (2010). *Merchants of Doubt: How a Handful of Scientists Obscured the Truth on Issues from Tobacco Smoke to Global Warming.* New York, NY: Bloomsbury.

Peeters, S., & Breedveld, B. (2009). Verkeerde inschatting van voedselrisico's. *Voeding Nu, 10,* 12–14.

Pidgeon, N., Kasperson, R. E., & Slovic, P. (eds.), (2003). *The Social Amplification of Risk.* Cambridge: Cambridge University Press.

Pidgeon, N., & Rogers-Hayden, T. (2007). Opening up nanotechnology dialogue with the publics: Risk communication or 'upstream engagement'? *Health Risk & Society, 9,* 191–210.

Rao, T. S .S., & Andrade, C. (2011). The MMR vaccine and autism: Sensation, refutation, retraction, and fraud. *Indian Journal of Psychiatry, 53,* 95–96.

Renn, O. (1991). Risk communication and the social amplification of risk. In R. E. Kasperson & P. M. J. Stallen (eds.), *Communicating Risks to the Public.* Dordrecht: Kluwer, pp. 287–324.

Renn, O. (2011). The social amplification/attenuation of risk framework: Application to climate change. *Wiley Interdisciplinary Reviews-Climate Change, 2,* 154–169.

Reynoso, H. M. (2003, October). *Under the volcano: what we should know about the Popocatépetl.* Panel contribution at the Association of Science and Technology Centers (ASTC) 2013 Annual Conference, Albuquerque, NM.

RIVM (2003). *Nuchter omgaan met risico's.* Bilthoven: RIVM.

Roeser, S., & Fahlquist, N. J. (2014). Moral emotions and risk communication. In J. Árvai & L. Rivers (eds.), *Effective Risk Communication.* London: Earthscan, pp. 204–219.

Sandman, P. M. (2012). *Responding to Community Outrage: Strategies for Effective Risk Communication.* Princeton, NJ: Peter M. Sandman.

Sandman, P. M. (2014). The Peter M. Sandman risk communication website. Retrieved August 27, 2018 from http://psandman.com.

Schlichting, I. (2013). Strategic framing of climate change by industry actors. A meta-analysis. *Environmental Communication, 7,* 493–511.

Semetko, H. A., & Valkenburg, P. M. (2000). Framing European politics: A content analysis of press and television news. *Journal of Communication, 50,* 93–109.

Silliman, B., & Wear, S. (2018). Conservation bias: What have we learned? In P. Karieva, M. Marcier, & B. Silliman (eds.), *Effective Conservation Science: Not a Dogma.* Oxford: Oxford University Press, p. 181.

Sjöberg, L., Moen, B. E., & Rundmo, T. (2004). *Explaining Risk Perception. An Evaluation of the Psychometric Paradigm in Risk Perception Research.* Trondheim: Rotunde.

Slovic, P. (2000). *The Perception of Risk.* London: Earthscan.

Timmer, M. (2010, October). *Vaccination against Cervical Cancer.* Paper presented at a Postdoctoral Course Risk Communication, Utrecht, NL.

Van Eeten, M., Noordegraaf-Eelens, L., Ferket, J., & Februari, M. (2012). *Waarom burgers risico's accepteren en waarom bestuurders dat niet zien.* Den Haag: Ministerie van Binnenlandse Zaken en Koninkrijksrelaties.

Van Woerkom, C. M. J. (1988). *Persuasieve voorlichting, een terreinverkenning ten behoeve van evaluatieonderzoek.* Wageningen: Landbouw Universiteit Wageningen.

World Health Organization (WHO) (2013). *Health and Environment: Communicating the Risk.* Copenhagen: WHO Regional Office for Europe.

Wynne, B. (1992). Misunderstood misunderstanding: Social identities and public uptake of science. *Public Understanding of Science, 1,* 281–304.

Chapter 8

Health Communication

Madelief G. B. C. Bertens, Joanne N. Leerlooijer,
and Maria E. Fernandez

8.1 Introduction

Although most people claim they value good health, they do not act accordingly. For example, people are usually aware of the negative consequences of smoking, drinking (too much) alcohol, having an unhealthy diet, or practicing unsafe sex, however, this awareness does not necessarily lead to healthy behavior (Kelly & Barker, 2016). To change these unhealthy practices, people must be motivated, have the appropriate skills, confidence, attitudes, and (social) support to facilitate behavior. Moreover, most health problems are multi-causal, influenced not only by behavioral factors but also by environmental conditions. Therefore, a comprehensive approach is needed to address these causes. Additionally, the determinants of behavior and environmental conditions are varied and include more than knowledge or awareness. They also include attitudes, skills, and self-efficacy related to the behavior. Thus, to be effective in influencing health, communication about health must do more than merely providing information.

Health is regarded as an essential and valuable imperative for well-being, and health communication is a way to enhance and stimulate healthy behavior. Health communication is based on behavioral, biomedical, psychological, and social sciences and applies scientific knowledge to promote health and well-being. Health communication as a profession is well-established and based on evidence-based approaches. The term 'health communication' is used as an umbrella term covering various communication practices in health education, health promotion, healthcare, biomedical communication, and patient education. Health communication draws on knowledge from many other fields, including epidemiology, life sciences, biomedical sciences, social and health psychology, communication sciences, science communication, marketing, sociology, and policy sciences.

While not the same, health communication and health education or health promotion have similar aims and approaches to planning. Health education entails 'any planned combination of learning experiences designed to predispose, enable, and reinforce voluntary behavior conducive to health in individuals, groups or communities' (Green & Kreuter, 1991, p. 432). Health promotion has a broader scope, ranging from disease prevention to wellness enhancement. It refers to 'any planned combination of educational, political, regulatory, and organizational support for actions and conditions of living conducive to the health of individuals, groups or communities' (Joint Committee on Health Education and Promotion Terminology, 2001, p. 101).

Health communication professionals are usually employed by the government, semi-public institutions, healthcare organizations, national knowledge institutes, and other health promotion institutions and organizations. They differ in expertise, qualifications, backgrounds, and degrees in fields such as health sciences, medicine, biology, food sciences, and health psychology with a specialization in health communication.

Health communication professionals plan, develop, implement, and evaluate interventions, such as public health campaigns or lifestyle programs. They conduct or commission research to better understand how messages and strategies can influence health-related behavior. They often co-develop programs or campaigns and identify how these can be implemented in the best way and can be involved in evaluation. Their health promotion strategies are systematically designed, implemented, and evaluated to ensure that objectives and communication strategies are relevant for the target audience, often called the priority group in health communication, and, thus, more likely to lead to intervention effectiveness. Health communication professionals use several explanatory and behavioral change models that play an important role in systematic theory-driven intervention planning.

This chapter introduces the field of health communication, its professionals, and how they work toward influencing health-related behaviors. As an introductory chapter, it only touches upon the basics of health communication practice and theory. It is by no means intended to be a complete overview of all of theories and practices underlying health promotion, education, and communication.

8.2 The Concept of Health

What is health? How does it relate to the body? Are people healthy when they are not ill? In 1948, the World Health Organization described health as 'a state of complete physical, mental and social well-being, and not merely the absence of disease or infirmity' (World Health Organization, 2006, p. 1). As such, health can be applied at a number of different levels as follows:

- *objective health*: health at an organic, biomedical level, as established by biomedical parameters or as diagnosed by a medical professional;
- *subjective health*: the perception of health or well-being experienced by the individual; and
- *social health*: health as determined by functioning in society.

Criticism of the WHO definition has resulted in a more dynamic definition based on the resilience or capacity to maintain and restore one's integrity, equilibrium, and sense of well-being: 'health as the ability to adapt and to self-manage' (Huber *et al.*, 2011). As such, someone with a physical impairment will not necessarily be deemed to be ill. Individuals with diabetes, for example, are able to function well socially and feel healthy if they are able to manage their disease and adjust their lifestyle.

The definition of health as the ability to adapt and to self-manage implies a different type of responsibility of the individual. In healthcare settings, this goes hand in hand with the increased demand for autonomy and shared decision-making by patients. The increased attention for participation of patients and the public has had implications for health communication. Health promotion involves cooperating and collaborating with the target population to understand their needs, identify the available resources, create

| Self-mobilization |
| Interactive participation |
| Functional participation |
| Participation for material incentives |
| Participation by consultation |
| Passive participation |
| Manipulative participation |

Figure 8.1:   A typology of participation (Pretty, 1995, p. 1252).

support, and mobilize the community. In fact, community participation is the basis of successful health promotion (South, 2014). Community members can participate on different levels. Participation ladders (e.g. Arnstein, 1969 and Pretty, 1995) visualize the extent to which citizens can participate in decision-making by defining rungs ranging from no, low- to high participation, or self-mobilization (Figure 8.1).

Increased participation and autonomy of citizens mean that they take responsibility for their own lives and environment. Similar processes and communication approaches can be found in the domains of environmental communication (see also Chapter 9). The (national) governmental agencies facilitate the process. Citizens are encouraged to set up health interventions themselves, and health workers are urged to enter into collaborative arrangements with organizations and citizens at different levels. In addition, an increasing number of interventions involve the use of methods and strategies designed to activate participants to make choices that will result in changed behavior. Self-regulation (Boekaerts, Pintrich & Zeidner, 2000) and motivational interviewing techniques (Miller & Rollnick, 2013) are rapidly growing in popularity.

There will always be a tension between full autonomy and civil mobilization on the one hand and authoritarianism and governmental control on the other hand (Van den Hoven & Kessler, 2011; see also Chapter 7). After all, health promotion is geared toward improving health of a group of people, influencing and changing behavior, or creating supportive environments. People are influenced to change their unhealthy or undesirable behavior, often to the benefit of the individual, and also for the greater good, for example, lower medical costs for society. The following are the important questions that need to be addressed: At what point does health promotion interfere with individual autonomy and does it become undesirable? How much limitation of personal autonomy is acceptable? Steps to deter people from driving under the influence of alcohol or drugs as a way of preventing fatal traffic accidents are considered very acceptable. Pressuring addicts to kick their habit and, by doing this, to avoid the nuisance they cause is accepted too. However, what is acceptable in efforts to address eating habits so as to prevent obesity?[1]

8.3 Health Promotion and Disease Prevention

The aim of health promotion and disease prevention is to influence health behaviors and environmental conditions to improve health and well-being. Prevention or disease management is often a goal of health promotion and health education on different levels (Baumann & Karel, 2013).

- *Primary prevention* refers to the prevention of health problems, diseases, and accidents before they occur, by affecting or removing risk factors.
- *Secondary prevention* includes the early diagnosis and treatment of health problems to moderate a disease, illness, or condition or to reduce the risk of reoccurrence.
- *Tertiary prevention* includes the care, treatment, and therapy to manage and limit the consequences of a condition or illness, in order to improve and/or maintain the quality of life.

Health education and health promotion often raise awareness, provide knowledge, and/or use persuasive communication by mass media campaigns. Even though providing knowledge or correcting misconceptions and addressing beliefs can sometimes be sufficient, behavior or lifestyle change generally requires more than just providing information. A combination of approaches such as information provision, persuasive communication

---

[1] See Buchanan (2006) for a discussion of 'paternalistic health information'.

strategies, confronting people with misconceptions, breaking habits, teaching people new skills, and putting certain facilities in place, are more likely to result in a change in behavior. For example, the most successful school-based nutrition education interventions include much more than an awareness campaign about unhealthy food and beverages (Murimi *et al.*, 2018). Successful interventions, such as the whole-school-based interventions, are intensive and comprehensive. They include promotion of a healthy school food environment and efforts to increase parental or family support (Wang & Stewart, 2013; Meiklejohn, Ryan & Palermo, 2016).

In 1974, lifestyle was identified and embraced as a major influence on health. It covers behaviors and practices such as eating habits and exercise patterns (Baum, 2008). Lifestyle includes someone's view of humankind and perspective of society: who and what you want to be, who and what you feel connected to, and the impact of these choices. Lifestyle includes, for example, the clothes one wears, the car one drives, the newspaper one reads, the friends one socializes with, the music one listens to, and the venues one visits. Lifestyle is developed reciprocally. In other words, although people choose to wear a certain outfit, this choice may be largely determined by their peers, parents, and other role models. The social and physical environment in which people grow up largely determines their lifestyle, whereas their lifestyle will also influence where they prefer to spend their social and physical life. Lifestyle, social position, and health are inextricably bound together (e.g. Feinstein, 1993). Needless to say, a health promoter must bear in mind the sociocultural context of individuals and groups and will need to consider lifestyle factors and preferences when designing communication strategies and resources. For example, in many countries, major health differences are present between various population groups. People with a lower socioeconomic status (SES), for example, are more likely to engage in unhealthy lifestyle behaviors, have less access to quality healthcare, and suffer more from physical and psychological health problems as compared to people with a higher SES (Pickett & Wilkinson, 2015).

It should be noted that in order to promote health, influencing people's health-related behavior is usually not sufficient. Health is influenced by the social, political, and environmental factors, a notion referred to as an ecological perspective to health. Health communication therefore increasingly encompasses multi-level communication strategies and interventions, such as tailored messages at the individual level, targeted messages at the group level, social marketing at the community level, media advocacy at the policy level, and media campaigns at the population level (Bernhardt, 2004; Golden *et al.*, 2015).

8.4 Communication and Health Policy

Health communication can be part of the implementation of a (public) health policy. Local and federal governments may employ various policy instruments to promote public health, including advocacy and communication (*nodality*), legislation and regulations (*authority*), financial incentives or sanctions (*treasure*), and by providing means to particular organizations who implement the health policy (*organization*) (Hood & Margetts, 2007). Often, a mix of policy instruments is used to influence people's behavior and their environment.

The first category, nodality, includes communicative instruments that can be used to facilitate a public debate, to inform the public, or to persuade people to adopt a healthy behavior, by increasing awareness, changing attitudes, promoting motivation, communicating favorable social norms, and informing people about changes in legislation and regulations and to bring facilities to their attention. Take, for instance, the example of smoking. Interventions can inform people about the negative consequences of smoking and also initiate a public debate about smoking on school grounds or playgrounds for children.

Another instrument that can be used is authority, that is, using legislative changes related to smoking such as prohibition (no-smoking laws) in public spaces. This could however be experienced as moralizing and patronizing, which could result in resistance (Lund, 2016). Coupling legislative changes with communication campaigns can reduce resistance and increase buy-in. For example, to realize smoke-free outdoor school grounds, it is important to combine legislation for outdoor smoking bans with collaboration, communication, and involvement of stakeholders at an early stage of the process (Rozema *et al.*, 2016).

The category treasure encompasses financial incentives or sanctions to guide behavior in the desired direction. In the smoking example, increasing the costs of cigarettes would be a financial instrument the government can use to influence people's behavior. Finally, organization relates to the means organizations have to implement policies. For example, the extent to which schools have sufficient resources and policies in place to create and maintain smoke-free schoolyards.

8.5 Health Communication Channels

Health communication messages can be delivered through a number of different channels, including mass media, the so-called small media, such as leaflets and brochures, and interpersonal communication. Mass media campaigns, which could run via television advertising and posters, for example, are well-known ways to convey messages across disciplines from advertising to health communication. Mass media campaigns can contribute to positive changes or prevent negative changes in health-related behaviors across large populations (Wakefield, Loken & Hornik, 2010). These campaigns appeared to be more successful for one-off or episodic behaviors, such as vaccination, than habitual behaviors such as food choices.

While these types of campaigns may increase knowledge and awareness, research shows that informational objectives are easier to accomplish by means of this type of campaign than attitudinal and behavioral objectives (e.g. Renes *et al.*, 2011). Moreover, several science communication studies show that increased knowledge rarely results in behavioral change (Noar, 2006; Snyder *et al.*, 2004). Despite evidence underscoring these limitations of mass media campaigns, mass media campaigns focusing on increasing knowledge remain popular in health communication (Renes *et al.*, 2011).

When it comes to raising awareness about health issues though, mass media campaigns can be important. They can have an agenda-setting function for the health issue concerned because of the large number of people the campaign may reach. An example is the increased awareness in colorectal cancer screening in the United States when a popular television personality talked about it and went through the process on television. In March 2000, following the tragic death of her husband from colon cancer at age 42, Katie Couric, an NBC anchor person, completed a colonoscopy live on the *Today Show*. This event was the start of a weeklong series promoting colon cancer awareness and promoting colorectal cancer screening.

One important disadvantage of mass media campaigns is the uncertainty about the communication process at hand. It is difficult to establish who has or has not received the message, let alone to ascertain how people will use the information if it does reach them (Koelen & Van den Ban, 2004). This problem is less of an issue in interpersonal communication.

In interpersonal communication, health communicators can more easily check through verbal and nonverbal interactions whether the priority group grasps the information. Forms of interpersonal communication are as follows:

- Face-to-face communication or dialogue, including patient education, counseling, personal instruction, or discussion. An example of face-to-face health communication consists of counseling sessions with a dietician aimed at changing nutrition habits.
- Lectures followed by an invitation to ask questions. Examples include meetings that are organized for pregnant women and their partners about their pregnancies, the birth itself, and taking care of their babies.
- Demonstrations, for example, a lesson on how to resuscitate someone or give a heart massage.
- Group information or information provided by one's own peers (this is also known as 'peer education') and self-help groups.

These forms of interpersonal communication reach fewer people, so they are relatively more expensive per person than mass media campaigns. However, participants are more engaged and have more opportunity to process information, consider arguments, and practice and develop skills. So, in that sense, they can be more effective. Interaction usually improves the impact on behavioral change (Koelen & Van den Ban, 2004).

8.6 Systematic Planning of Health Communication Interventions

Communication professionals need to ensure that communication strategies and methods meet the objectives of the communication effort, are appropriate for the priority group, and are theory and evidence based (see also Chapters 3 and 9). Intervention Mapping (IM) is a protocol for planning systematic theory- and evidence-based interventions in six steps (Bartholomew-Eldredge *et al.*, 2016). IM uses an ecological and systems approach to health, meaning that an individual's larger social context is taken into account. IM also encourages collaboration between the various stakeholders through the participation of the priority group and other stakeholders, including intervention developers, implementers, and policy-makers in the development process. Table 8.1 shows the six steps of the iterative process of intervention mapping.

An example of an intervention developed using the IM framework is Por Nuestros Hijos (For our Children). The intervention targeted Latino parents of low-income or underinsured girls and boys aged 11–17 years old with an aim to increase HPV vaccination among their daughters (Rodriguez *et al.*, 2018).

Step 1: Problem analysis (logic model of the problem)
The development of Por Nuestros Hijos began with a needs assessment to understand the local burden of HPV in low-income or underinsured Texans as well as factors influencing parental decisions to vaccinate, including attitudes about the vaccine, beliefs, and barriers. The needs assessment included analyzing literature, assembling a community action board, and conducting qualitative research with members of the target population. The needs assessment identified several determinants associated with parental vaccination acceptance, including knowledge about HPV, attitudes toward the vaccine, and perceived barriers.

Step 2: Intervention objectives (logic model of change)
In this step, planners list performance objectives (or sub-behaviors) that describe what the parents have to do to vaccinate their daughters and include communication with a healthcare provider, agreeing to the vaccine for their child, and obtaining subsequent vaccine doses. In this step, the planner also identifies determinants of vaccination, such as, 'why would parents vaccinate their child?', and creates matrices of change objectives by crossing determinants and performance objectives. These matrices then drive the selection of methods and practical applications as well as content and messages of the intervention.

Table 8.1: IM steps and tasks.

Step 1: Logic model of the problem	• Establish and work with a planning group. • Conduct a needs assessment to create a logic model of the problem. • Describe the context for the intervention including the population, setting, and community. • State program goals.
Step 2: Program outcomes and objectives — logic model of Change	• State expected outcomes for behavior and environment. • Specify performance objectives (sub-behaviors) for behavioral and environmental outcomes. • Select determinants for behavioral and environmental outcomes. • Construct matrices of change objectives. • Create a logic model of change.
Step 3: Program design	• Generate program themes, components, scope, and sequence. • Choose theory- and evidence-based change methods. • Select or design practical applications to deliver change methods.
Step 4: Program production	• Refine program structure and organization. • Prepare plans for program materials. • Draft messages, materials, and protocols. • Pre-test, refine, and produce materials.
Step 5: Program implementation plan	• Identify potential program users (implementers, adopters, and maintainers). • State outcomes and performance objectives (sub-behaviors) for program use. • Construct matrices of change objectives for program use. • Design implementation interventions.
Step 6: Evaluation plan	• Write effect and process evaluation questions. • Develop indicators and measures for assessment. • Specify the evaluation design. • Complete the evaluation plan.

Evaluation

Implementation

Source: Adapted from Bartholomew-Eldredge *et al.* (2016, p. 13).

Step 3: Methods (program design)

The communication strategy for Por Nuestros Hijos includes a photo novella and tailored interactive multimedia interventions, in the form of videos and interactive educational modules on a tablet, to personalize health messages and allow each parent in the intervention to receive messages focused on specific questions and determinants. The overarching theme of the intervention is protection: the protection the HPV vaccine gives as well as the protection parents can give their child through HPV vaccination. Negative consequences of HPV are used to create anticipatory regret, and modeling of behaviors is used to show a mother agreeing to vaccinate her child against HPV.

Step 4: Program

The program includes an interactive multimedia online intervention, which incorporates theoretical methods such as modeling and persuasion and uses an entertainment education approach to deliver the messages including a story about a parent making the decision to vaccinate their child. The delivery channels include interpersonal (by a lay health worker), small media (tailored interactive program on tablets), and a photo novella.

Step 5: Implementation

IM was used to develop objectives for lay health workers, who deliver Por Nuestros Hijos to parents of adolescent Latinas. A lay health worker organization was included as a member of the community advisory board. Lay health workers were recruited through this organization and trained over a 2-day period. In the training, medical experts taught the lay health workers about HPV and the vaccine, staff trained them in skills needed to implement the intervention, and lay health workers practiced with delivering the intervention.

Step 6: Evaluation

Program evaluation focuses on the impact of the intervention on parental acceptance of the vaccine for their children. Preliminary results indicate that parents receiving the intervention are more likely to vaccinate their children against HPV than parents who were not exposed to the intervention. In the project, IM was useful for determining research questions related to the intermediate impact of the program (e.g. influence on determinants) and for designing process evaluation measures that corresponded with the implementation plan.

8.7 Theories and Models in Health Communication

Before developing health communication messages, it is important for health communication professionals to understand which factors or behavioral determinants influence the behavior or environmental condition that is targeted, as well as how people process information. With this in mind, professionals draw on various commonly used sociopsychological theories or models to either explain behavior and environmental conditions or to influence them. The overview in this section features the most applied or used theories and models and does not aim to be complete. For a more elaborate overview, see, for example, Glanz, Rimer & Lewis (2005) and Bartholomew-Eldredge *et al.* (2016). In addition, empirical evidence or criticism of the various concepts, theories, and models is not included in this chapter for the lack sake of space.

8.7.1 *How to Explain Behavior*

Commonly used theories or models that are used to understand and explain behavior include the Social Cognitive Theory (Bandura, 1986) and the so-called theories of planned or reasoned behavior (Fishbein & Ajzen, 1975; Ajzen, 1991).

According to the Theory of Planned Behavior (TPB), *intention* is the most important predictor of behavior. Intentions guide behavior. An example of an intention in relation to safe sex: 'I will have safe sex with new or casual partners' or 'I will not use condoms when having sex with my partner'. Intentions indicate intensity, the amount of effort someone will exert: 'I want to be able to have safe sex with *all* of my casual partners and I intend to *always* have a condom with me'. *Attitude, social norms*, and *self-efficacy* are the underlying determinants that explain intention. *Skills* help someone to act on intentions (e.g. assertiveness skills), whereas *barriers* hinder someone from doing this (e.g. condoms are scarcely available).

- *Attitude* is the belief that someone has toward his or her behavior and the expectation of what will ensue from this behavior. The above involves an assessment of the behavior. Assessments regarding safe sex are for example: 'I find condoms really awkward to use', or 'Sex with a condom is a lot less enjoyable'. The attitude and outcome expectations that someone has will be determined in part by knowledge and previous experiences.
- *Social norms*, also referred to as subjective norms or social influence, represent what someone believes important others would do. 'None of my friends use condoms. They think condoms are awkward too'. And what someone thinks that other, important others think he or she ought to do and the approval or disapproval of certain behavior. 'My mother thinks that I should always have safe sex'. It is the normative beliefs weighted by the motivation to comply. Even though a mother is an influential other, teenagers will often be more inclined to do what their peers feel is important. The perceived social support or perceived social pressure form part of social influence as well.
- *Self-efficacy* involves an assessment of one's own possibilities to engage in the behavior. Factors like perceived difficulty and confidence in one's own ability play a role here; the extent to which individuals themselves or others are able to change something. For example, the confidence one has to continue using a condom.

TPB is a general theory that is used to explain a wide range of behaviors, not only health-related ones. For instance, it is also often used for pro-environment behaviors (see Chapter 9).

In comparison to TPB, some theories and models have a stronger focus on weighing the benefits and losses of expected outcomes. These include the Health Belief Model (HBM), the Protection Motivation Theory (PMT), and the Extended Parallel Process Model (EPPM).

Health Belief Model

HBM is frequently used to understand health behavior and provides a basis for designing interventions. The model was developed in the 1950s (Becker, 1974) and is based on the notion that the decision to engage in healthy behavior is determined by a person's perceptions of personal susceptibility to, and the severity of, a particular condition or illness balanced against perceived benefits and barriers. These determinants influence people's readiness to act. An added concept, cues to action, activates that readiness and stimulates overt behavior. Another addition to the HBM is the concept of self-efficacy, the same determinant that is also included in TPB and Social Cognitive theory (Rosenstock, Strecher & Becker, 1988).

Protection Motivation Theory
PMT (Rogers, 1975) is also based on the principle that people are inclined to protect themselves and their health. Two cognitive processes kick in when a health threat arises: threat appraisal and coping appraisal. On the one hand, people assess the personal threat. They weigh the perceived severity of and personal susceptibility to the illness or complaint in question. The greater they perceive a threat, the greater the motivation to protect oneself. On the other hand, people assess possibilities that will enable them to cope with the threat, the effectiveness of potential responses (response efficacy), and one's ability to undertake action (self-efficacy). Together, these appraisals will generate protection motivation — an intention to self-preservation behavior (Kok *et al.*, 2014).

Extended Parallel Process Model
EPPM (Witte, 1998) shows that the motivation to ensure one's self-preservation may take on different forms depending on the outcome of threat and coping appraisal. If people do not perceive any threat or fear, they will not respond at all. They will not undertake any action. If people experience too much threat or fear but they are not convinced of the effectiveness of an alternative behavior and/or are not confident of their self-efficacy, the resulting response may be defensive, oriented to avoidance. They deny the risks, they will not take the message seriously or they will disregard the messenger or message. If people do perceive a threat and know how to cope with the threat and are confident, they will respond effectively by addressing the threat: this is referred to as protection motivation. The EPPM explains how fear-arousing messages can result in adverse responses. Fear arousal can only be effective if people feel confident they can do something to prevent the threat (Kok *et al.*, 2014).

8.7.2 *How to Influence Behavior*

Being able to understand and explain behavior aids the development of health communication interventions that aim to influence behavior. Influence can be exerted on the determinants of behavior, including attitudes, social norms, and self-efficacy. Knowledge is an essential element of most behavioral determinants. Given this fact, knowledge transfer is usually the first step in communication, but as stated earlier, knowledge alone is usually not enough to change behavior. Smokers know that smoking is bad for their health, but will still light up a cigarette.

For example, the Por Nuestros Hijos intervention (see Section 8.6) aimed among others at increasing the mother's self-efficacy to support her daughter to vaccinate against HPV. Portraying role models, including mothers depicted in the photo novella in Por Nuestros Hijos, are used to shift self-efficacy. People are able to learn through observation, so by having a role model display the behavior and skills and observing the behavior being reinforced, self-efficacy increases.

It is important to note that many of the social psychological theories discussed in this section are based on cognitive processes. The pros and cons of displaying certain behavior are often weighed, consciously or subconsciously. This process is addressed and elaborated when designing health promotion interventions. Nevertheless, people also make many (health) behavioral decisions without consciously or deliberately considering these factors. People are all creatures of habit and primarily do what feels comfortable and what will cost the least effort. *Dual-process models* (e.g. Kahneman & Frederick, 2002) state that there are two different modes of processing and decision-making: an implicit, unconscious, automatic, fast process (referred to as 'system 1') and an explicit, deliberate, conscious, and slow process (referred to as 'system 2').

The *Elaboration Likelihood Model* (ELM) (Petty, Barden & Wheeler, 2002) integrates the conscious and subconscious decision-making processes. The model states that people who are engaged with the subject, have the ability to process the message, and are motivated to process the message will do so via the so-called central route. In other words, they will consciously consider the message and weigh up the various pros and cons based on the validity of arguments. This central processing results in a stable change in attitude, both positively accepting the message and negatively rejecting the message. People who are not engaged in the subject will not have any interest in the message and will process the information via the peripheral route, paying attention to the superficial characteristics of the information provider, such as the look and appeal of the messenger or message.

The goal in health communication is for the priority group to process information centrally. A way to achieve this is by engaging the priority group and increasing personal relevance. Consider, for example, a role-play. The peripheral route of information processing can be used to appeal to the audience and bring attention to the subject and, by doing this, increase the engagement and chances of central processing (Brug, Van Assema & Lechner, 2012).

Edutainment, an approach in which the information message is embedded in the storyline of a video or television program, draws on ideas from the ELM to ensure that the message is delivered to an unengaged priority group. This makes it possible to bring a topic to the attention of a non-involved priority group. Because the viewers project themselves into the situation in question, they become involved and are able to process the message.

8.8 Evaluation and Effectiveness of Health Communication

Many factors can influence the effectiveness of health communication programs (Sixsmith *et al.*, 2014). Certainly, systematic planning to identify important behavioral and environmental factors influencing the health problem are important factors. Health communication interventions that are carefully planned considering these factors, and selecting messaging and delivery strategies that will affect them, will increase the likelihood that the intervention will be effective. Stating specific behavioral goals including the desired outcome, specific behavior, and target level of change so that outcomes are measurable, are other factors.

Additionally, considering demographic characteristics of the priority group is essential. This may include a need to focus on vulnerable populations and a consideration of cultural or other factors such as varying social norms, perceived risk, misinformation, beliefs, attitudes, and potential barriers that may influence behavior (Snyder, 2007).

Another important factor influencing effectiveness of health communication campaigns is appropriate selection of communication activities and channels. These should be carefully considered and analyzed within the context of influencing decision-making among the target population(s). It is helpful to solicit participation of priority group members and community organizations in the design and implementation of campaigns and to establish community ties to inform not only content but also implementation and dissemination strategies.

As previously mentioned, message content and presentation are critical. Specific messages should be designed for each specific priority group. Additionally, planners should consider priority group reach. Impact of the intervention is a function of not only its effectiveness but also the extent to which it reaches the priority group. Thus, selection of channels that will increase the reach should be considered. Creativity of message development is also an important component of what makes health communication campaigns effective. Messages should capture attention of the priority group. This can be done in unique ways, such as creating culturally specific pictures, logos, and slogans (e.g. Snyder, 2007).

Various studies show that health communication interventions have small effects on health-related behaviors such as smoking, alcohol reduction, sexual behavior, or cancer screening (Snyder *et al.*, 2004; Anker *et al.*, 2016). One of the reasons is that these interventions have relied primarily on communication and education to influence people's behavior, but did not sufficiently take social determinants of health into account.

Social determinants include people's environmental, socioeconomic, and cultural settings. These are less recognized and addressed in health communication interventions, whereas they relate to the root causes of health problems. This is particularly the case among vulnerable groups who are socially or economically disadvantaged and have a higher risk of chronic conditions such as cardiovascular diseases, cancer, and diabetes type 2. Nutbeam (2000) introduced the concept of *health literacy*, whereby health education is directed toward equipping people to overcome structural barriers to health. Examples include aiming at skills development to improve people's access to health information and their capacity to use it effectively.

8.9 Conclusion

Health communication is an essential element of health education, health promotion, biomedical communication, patient information, and genetic counseling. Health is influenced by both individual behaviors and factors in people's social and physical environment. In the past, many health promotion messages and strategies solely relied on cognitive rational models and emphasized knowledge transfer. Important considerations of other influences on health behavior such as habit and environmental influence are critical to improve the effectiveness of these messages and strategies (Sheeran & Webb, 2016). Increasingly, health promotion interventions focus on changing the physical environment or amending legislation and regulations.

In health communication, providing information is generally not sufficient to influence people's health-related behaviors and habits. In many situations, it is also necessary to change people's attitudes, their social norms, and self-efficacy. In addition to these cognitive factors, there is increasing attention to influence people's less conscious system such as emotions and automatic behaviors.

Health communication should therefore take into account a focus on both behavior and environment, on knowledge and other determinants of behavior, and on the conscious and unconscious factors that influence people's behaviors. The potential for having an effect on health also depends heavily on the care with which health communication interventions are developed and delivered. Systematic processes such as IM can guide the careful development of programs through steps that consider the problem at multiple levels, the determinants (based on theory and evidence), and the delivery.

References

Ajzen, I. (1991). The theory of planned behaviour. *Organisational Behaviour and Human Decision Processes, 50,* 79–211.

Anker, A. E., Feeley, T. H., McCracken, B., & Lagoe, C. A. (2016). Measuring the effectiveness of mass-mediated health campaigns through meta-analysis. *Journal of Health Communication, 21*(4), 439–456.

Arnstein, S. R. (1969). A ladder of citizen participation. *Journal of the American Planning Association, 35*(4), 216–224.

Bandura, A. (1986). *Social Foundations of Thought and Action.* Englewood Cliffs, NJ: Prentice Hall.

Bartholomew-Eldredge, L. K., Markham, C. M., Ruiter, R. A. C., Fernández, M. E., Kok, G., & Parcel, G. S. (eds.). (2016). *Planning Health Promotion Programs: An Intervention Mapping Approach* (4th edn.). San Francisco, CA: Jossey-Bass.

Baum, F. (2008). *The New Public Health.* Oxford: University Press.

Baumann L. C., & Karel A. (2013). Prevention: primary, secondary, tertiary. In M. D. Gellman, & J. R. Turner (eds.), *Encyclopedia of Behavioral Medicine.* New York: Springer, pp. 99–136.

Becker, M. H. (1974). The Health Belief Model and personal health behavior. *Health Education Monographs, 2,* 409–419.

Bernhardt, J. M. (2004). Communication at the core of effective public health. *American Journal of Public Health, 94*(12), 2051–2052.

Boekaerts, M., Pintrich, P. R., & Zeidner, M. (eds.) (2000). *Handbook of Self-regulation.* San Diego, CA: Academic Press.

Brug, J., Van Assema, P., & Lechner, L. (eds.) (2012). *Gezondheidsvoorlichting en gedragsverandering. Een planmatige aanpak.* Assen: Van Gorcum.

Buchanan, D. R. (2006) Autonomy, paternalism, and justice: Ethical priorities in public health. *American Journal of Public Health, 98*(1), 15–21.

Feinstein, J. S. (1993). The relationship between socioeconomic status and health: A review of the literature. *Milbank Quarterly, 71*(2), 279–322.

Fishbein, M., & Ajzen, I. (1975). *Belief, Attitude, Intention, and Behavior.* New York: Wiley.

Glanz, K., Rimer, B. K., & Lewis, F. M. (eds.), (2005). *Health Behavior and Health Education. Theory, Research and Practice.* San Francisco: Jossey-Bass.

Golden, S. D., McLeroy, K. R., Green, L. W., Earp, J. A., & Lieberman, L. D. (2015). Upending the social ecological model to guide health promotion efforts toward policy and environmental change. *Health Education & Behavior, 42*(1S), 8S–14S.

Green, L. W., & Kreuter, M. W. (1991). *Health Promotion Planning. An Educational and Environmental Approach.* Mountain View: Mayfield Publishing Company.

Hood, C. C., & Margetts, H. Z. (2007). *The Tools of Government in the Digital Age.* Palgrave Macmillan International Higher Education.

Huber, M., Knottnerus, J. A., Green, L., van der Horst, H., Jadad, A. R., Kromhout, D., Leonard, B., Lorig, K., Loureiro, M. I., van der Meer, J. W., Schnabel, P.,

Smith, R., van Weel, C., & Smid, H. (2011). How should we define health? *British Medical Journal, 343*, d4163.

Joint Committee on Health Education and Promotion Terminology (2001). Report of the 2011 Joint Committee on Health Education and Promotion Terminology. *American Journal of Health Education, 43*(2), 1–19.

Kahneman, D., & Frederick, S. (2002). Representativeness revisited: Attribute substitution in intuitive judgement. In T. Gilovich, D. Griffin, & D. Kahneman (eds.), *Heuristics and Biases: The Psychology of Intuitive Judgment.* Cambridge, UK: Cambridge University Press, pp. 49–81.

Kelly, M. P., & Barker, M. (2016). Why is changing health-related behaviour so difficult? *Public Health, 136*, 109–116.

Koelen, M. A. & van den Ban, A. W. (2004). *Health Education and Health Promotion.* Wageningen: Wageningen Academic Publishers.

Kok G., Bartholomew L. K., Parcel G. S., Gottlieb N. H., & Fernández M. E. (2014). Finding theory- and evidence-based alternatives to fear appeals: Intervention Mapping. *International Journal of Psychology, 49*(2), 98–107.

Lund, M. (2016). Exploring smokers' opposition to proposed tobacco control strategies. *Nordic Studies on Alcohol and Drugs, 33*(4), 321–334.

Meiklejohn S., Ryan L., & Palermo C. (2016). A systematic review of the impact of multi-strategy nutrition education programs on health and nutrition of adolescents. *Journal of Nutrition Education and Behavior, 48*(9), 631–646.

Miller, W. R., & Rollnick, S. (2013). *Motivational Interviewing: Preparing People for Change* (3rd edn.). New York, NY: Guilford Press.

Murimi, M. W., Moyeda-Carabaza, A. F., Nguyen, B., Saha, S., Amin, R., & Njike, V. (2018). Factors that contribute to effective nutrition education interventions in children: A systematic review. *Nutrition Reviews, 76*(8), 553–580.

Noar, S. M. (2006). A 10-year retrospective of research in health mass media campaigns: Where do we go from here? *Journal of Health Communication, 11*(1), 21–42.

Nutbeam, D. (2000). Health literacy as a public health goal: A challenge for contemporary health education and communication strategies into the 21st century. *Health Promotion International, 15*(3), 259–267.

Petty, R. E., Barden, J., & Wheeler, S. C. (2002). The elaboration likelihood model of persuasion: Health promotions that yield sustained behavioral change. In R. J. DiClemente, R. A. Crosby, & M. C. Kegler (eds.), *Emerging Theories in Health Promotion Practice and Research.* San Francisco: Jossey-Bass, pp. 71–99.

Pickett, K. E., & Wilkinson, R. G. (2015). Income inequality and health: A causal review. *Social Science & Medicine, 128,* 316–326.

Pretty, J. (1995). Participatory learning for sustainable agriculture. *World Development, 23*(8), 1247–1263.

Renes, R. J., van de Putte, B., van Breukelen, R., Loef, J., Otte, M., & Wennekes, C. (2011). *Gedragsverandering via campagnes.* The Hague: Ministry of General Affairs, Public Information and Communication Service.

Rodriguez, S. A., Roncancio, A. M., Savas, L. S., Lopez, D. M., Vernon, S. W., & Fernández, M. E. (2018). Using Intervention Mapping to develop and adapt two educational interventions for parents to increase HPV vaccination among Hispanic adolescents. *Frontiers in Public Health, 6,* 164.

Rogers, R. W. (1975). A protection motivation theory of fear appeals and attitude change. *Journal of Psychology, 91,* 93–114.

Rosenstock, I. M., Strecher, V. J., & Becker, M. H. (1988). Social learning theory and the health belief model. *Health Education Quarterly, 15*(2), 175–183.

Rozema, A. D., Mathijssen, J. J. P., Jansen, M. W. J., & van Oers, J. A. M. (2016). Schools as smoke-free zones? Barriers and facilitators to the adoption of outdoor school ground smoking bans at secondary schools. *Tobacco Induced Diseases, 14*(1), 10.

Sheeran, P., & Webb, T. L. (2016). The intention–behavior gap. *Social and Personality Psychology Compass, 10*(9), 503–518.

Sixsmith, J., Fox, K-A., Doyle, P., & Barry, M. M. (2014). *A Literature Review on Health Communication Campaign Evaluation with Regard to the Prevention and Control of Communicable Diseases in Europe.* Stockholm: European Center for Disease Control.

Snyder, L. B., Hamilton, M. A., Mitchell, E. W., Kiwanuka-Tondo, J., Fleming-Milici, F., & Proctor, D. (2004). A meta-analysis of the effect of mediated health communication campaigns on behaviour change in the United States. *Journal of Health Communication, 9*(S1), 71–96.

Snyder, L. B. (2007). Health communication campaigns and their impact on behavior. *Journal of Nutrition Education and Behavior, 39*(2), S32–S40.

South, J. (2014). Health promotion by communities and in communities: Current issues for research and practice. *Scandinavian Journal of Public Health, 42*(Suppl 15), 82–87.

Van den Hoven, M., & Kessler, C. (2011). *Preventie en ethiek.* The Hague: Boom Lemma.

Wakefield, M. A., Loken, B., & Hornik. R. C. (2010). Use of mass media campaigns to change health behaviour. *Lancet, 376*, 1261–1271.

Wang, D., & Stewart, D. (2013). The implementation and effectiveness of school-based nutrition promotion programmes using a health-promoting schools approach: A systematic review. *Public Health Nutrition, 16*(6), 1082–1100.

Witte, K. (1998). Fear as motivator, fear as inhibitor: Using the extended parallel process model to explain fear appeal successes and failures. In P. A. Andersen, & L. K. Guerrero (eds.), *Handbook of Communication and Emotion: Research, Theory, Applications, and Contexts*. San Diego, CA: Academic Press, pp. 423–450.

World Health Organization (2006). *Constitution of the World Health Organization — Basic Documents* (45th edn.), Supplement, October 2006, p. 1.

Chapter 9

Environmental Communication*

Liesbeth de Bakker and Eric A. Jensen

9.1 Introduction

Communicating about the environment has received increasing attention over the past decades and takes place in society in many different settings. For example, communication about the issue of climate change has become a major global challenge. Through the news, entertainment media, in schools, at local zoos, in corporate policies and many other aspects of contemporary life, topics such as the local environment, nature areas, healthy living infrastructure, the changing climate, and sustainable behavior are being actively discussed and debated.

The International Environmental Communication Association (IECA) defines environmental communication as follows:

> Environmental communication [addresses...] environmental issues and problems, and our relationship to the rest of nature. [...] Anyone who is participating in these discussions is engaging in the activity of environmental communication. That includes everyone from the most passionate environmental advocates, to the fiercest opponents of ecological protections. In this sense, it is both a lay activity that anyone can undertake, and a field of practice that professional communicators have created (IECA, 2019).

* The authors would like to thank Dr. F. Weder (Alpen-Adria-Universität Klagenfurt) and Dr. R. van den Born (Radboud University Nijmegen) for feedback and ideas on this chapter.

The IECA highlights that for many, a key defining feature of environmental communication is its status as a crisis field of research and practice. This means that good communication about the environment 'is essential if we are to avoid violent conflicts and address environmental health and justice issues in the most effective ways possible. Accordingly, a central goal of the field is to discern and promote good practices' (IECA, 2019).

This field is fueled by a sense of urgency stemming from large-scale environmental dangers such as those posed by climate change and the ongoing mass extinction of wildlife species caused by human actions. This stark global reality draws people and resources to this field, while creating pressure for rapid action.

Environmental communication is not new, however. For millennia, people have talked about and developed solutions for environmental challenges, ranging from water pollution to traffic. Moreover, the question of how humans can live in balance with their natural environment has been a hot topic in many places for hundreds of years. Some public health-oriented reforms have been strongly linked to environmental risks, dating back to Ancient Roman times when aqueducts drew from more distant lakes and rivers to provide fresh drinking water to polluted urban centers. Some key developments in the recognition of — and communication about — environmental risk can be identified more recently, for example, with movements developing in the United States at the beginning of the 1900s protecting specific areas and species. For example, in the case of the American buffalo, the species suffered a drastic decline from up to 30 million individual animals at one point to less than 1,000 in the wild in the 1800s. The American Bison Society formed in 1905, helping to successfully save the species from extinction. Moreover, in the United States as well, the Save the Redwoods League was founded in 1918, initially helping to establish protected areas for redwood trees in the 1920s. A growing public acknowledgement that natural resources are inherently limited was linked to such developments in environmental preservation and wildlife conservation.

In the 1960s and 1970s, renewed worries about the impact of industrialization on the environment surged in the developed world. A widely cited example, Rachel Carson's book *Silent Spring*, brought public attention to the dangers of pesticides in the early 1960s. Public and government interest in problems associated with pollution and the wider environmental consequences of industrialization increased in this period. Moreover, infamous accidents at nuclear energy power plants in both Russia (Chernobyl accident in the then USSR) and the United States (Three Mile Island accident) helped to drive the expansion of environmental (including anti-nuclear) organizations. This helped ignite a growing environmental awareness and broader acceptance of the need for more environmental protection and action in the late 20th century.

Then, at the dawn of the 21st century, climate change became one of the main foci for environmental communication, having been an increasing concern in the last decade of the 20th century. This issue is now drawing attention and resources as the risk of climate change increasingly becomes a visible, daily reality that directly affects millions of people each year through rising sea levels, forest fires, more severe weather patterns, and other dangers.

Environmental sustainability has become a key term in the context of environmental communication, reflecting the focus on making a positive change in humans' relationship to the environment. This concept refers to the long-term imperative to reduce human demands on the Earth's natural resources to 'ensure that [the world] meets the needs of the present without compromising the ability of future generations to meet their own needs' (World Commission on Environment and Development, 1987, p.1).

The idea of environmental sustainability is often combined with the need for international economic development to keep raising living standards under the banner of *sustainable development*. Decades of international work to develop policy and practice in this domain of sustainable development has led to the current United Nations (UN)–defined focus on 17 Sustainable Development Goals (or SDGs). Formally adopted by the UN General Assembly in 2015, the 17 SDGs present an overview of the world's most important social and economic problems such as income inequality and poverty reduction with environmental sustainability concerns. The notion of sustainability in general, and the SDGs in particular, offers a positive way forward by focusing on a vision of a world that lives both with a high standard of living and limited impact on the Earth's resources.

As a field of practice, environmental communication is growing and developing fast. Environment-related problems increase in number, size, impact, and complexity every year. This increasing complexity of environmental problems brings on a particular challenge. These days what scientists and other experts say is often called into question. Trust, or the lack thereof, in knowledge and messengers plays a very important role. Credibility is at stake. This has a huge impact on the quality and effectiveness of communication about environmental issues and demands more sophisticated, more interactive, engaged approaches and tools to support the ultimate goal of developing a more sustainable society. Even though this chapter does not explicitly deal with psychology, sociology, marketing, and policy, these fields strongly influence how environmental communication is understood and practiced, and aspects of these fields are highlighted throughout this chapter.

Although it has a distinct focus, environmental communication has a lot in common with the field of science communication. Not only do aspects of science, such as the study of nature and environment, comprise part of the scope of environmental communication but also science and technology are sources of methods and tools to deal with environmental challenges. This means that how people view science can affect their interpretations of environmental problems (see Chapter 2).

This chapter describes what environmental communication looks like in practice, key contexts in which it operates, and theoretical models that are driving its development. It also identifies approaches and tools commonly used to address challenges in environmental communication.

9.2 Environmental Communication in Practice

Environmental communication has specific goals and a great variety of stakeholders and publics. It often aims to engage with how people live and the effects of such lifestyles on natural resources. Human-made problems to do with the environment include global warming, the depletion of the Earth's resources, waste and pollution, environmental degradation, endangered species, biodiversity, invasive species, and more. Environmental communication can focus on engaging citizens about their role in such issues, encouraging children or young people to acquire and foster specific values such as caring for the environment or training professionals to develop more sustainable practices.

In terms of stakeholders and publics, many actors are involved in environmental communication processes, from government agencies to international expert panels such as the UN Intergovernmental Panels on

Climate Change (IPCC) and Biodiversity and Ecosystem Services. Also, media organizations, large and small companies, and local non-governmental organizations (NGOs) dealing with issues relating to the environment in specific communities participate. Each actor is responsible for developing its own communication approach based on its goals and an assessment of its specific target audiences. Important categories of environmental communicators include the following: scientific experts; journalists and the media; NGOs; nature and environment centers; public relations and marketing professionals; and policy-makers.

9.2.1 *Scientific Experts*

When it comes to scientific experts, Irwin, Bucchi, Felt, Smallman & Yearley (2018, p. 12) note: 'What counts as credible evidence and as trustworthy "public proof" is central when it comes to communication and engagement with environmental issues. This is in turn closely connected to questions of who is regarded as holding relevant expertise, how diverse actors claim authority over problem definitions, and whose values, concerns and vulnerabilities should matter when it comes to societal choices.'

Scientists have long played a key role in environmental communication as academic experts disseminating research findings to public and policy audiences, as well as in the role of public intellectuals offering their analysis of the challenges and possible solutions. Scientists often participate as experts on specific topics within governmental panels focused on addressing environmental problems, such as rising sea levels due to climate change. Some scientists participate in public engagement and take part in public debates. While natural scientists have long been the main source of expertise in public discussions about the environment, there is increasing recognition of the diversity of social science expertise that is needed to address the root causes of many environmental problems (Bennett *et al.*, 2017).

9.2.2 *Journalists and the Media*

Journalists have always fulfilled different roles. The cheerleader role is the one where they bring the good news, for instance, about new innovative techniques to deal with environmental problems, such as the story of Boyan Slat, a young Dutch researcher and entrepreneur, who is trying to scoop the plastic waste out of the world's oceans. Another important role is that of watchdog, in which they signal and analyze environmental problems, such as fossil fuel depletion, and urge government or others responsible to do something about it (see Chapter 6).

During the late 20th century, in the eyes of those in power, there was poor (uninformed) and sensationalist media coverage of environmental problems, think, for instance, about genetically modified (GM) crops and the related potential environmental risks (Smallman, 2014). This also shed a bad light on the people involved, the biotechnologists, GM crop researchers, and related institutes (Bucchi & Trench, 2014). But, just through informing better, this perceived problem could not be solved, so gradually a call emerged for a more inclusive approach, taking public perspectives, values, and emotions more into account.

It is important to note that large transformations of the media landscape are currently taking place, with the loss of influence of the traditional communication forms, such as television and printed newspapers, and the rise of social media platforms, such as YouTube and Facebook. This allows many people, both laypersons and experts, to become journalists themselves (Horst, Davies & Irwin, 2017). Indeed, many sustainability blogs — such as Going zero waste and Simple solutions for natural living — are gaining a substantial public following by giving tips for a more sustainable lifestyle in terms of waste reduction, eco-friendly traveling and clothing, and energy conservation schemes. They try to provide a constructive, positive outlook on a future riddled with environmental challenges. Through their personal webpages and channels, bloggers and vloggers engage in informal discussions about environmental issues and preferences.

9.2.3 Non-governmental Organizations

Within environmental communication, NGOs have an important role to play. They are non-commercial and non-government funded but serve the society and its (environmental) needs. Depending on their aims and focus, either nature or environment, they can play a more education-, conservation-, or policy-oriented role.

Key players in the international environmental NGO scene are the World Wildlife Fund and Greenpeace. The World Wildlife Fund is focused on nature and conservation and Greenpeace is more all-encompassing, focusing on protecting the environment and stimulating pro-environmental behavior. Greenpeace often works through public campaigns and actions and has both paid and volunteer activists who go to hotspots to protest against, for instance, the transport of nuclear waste, or the building of new fossil fuel-based power plants. In addition, the NGO supports local initiatives, empowers individuals who want to focus on regional problems, by appealing to them via texts on the homepage of their website: 'Calling all planet lovers with a wish to 'save the world' to team up and join forces to stand up to those in power to make the dream of a green and just world a reality' (Greenpeace, 2018).

9.2.4 Nature and Environment Centers

Often, major cities around the world have local city gardens, science museums and centers, and zoos or aquariums that talk about the environment with their visitors. These organizations are involved in environmental communication that makes use of informal science education approaches (see Chapter 5). Whereas at the end of the 20th century their work mainly focused on environmental pollution, waste reduction, and nature protection, now, also here environmental conservation and sustainability are an important focus in their public communication.

By providing visitors, mainly children, with experiences with nature or representations of nature, staff members hope to motivate individuals to start taking more care for nature, and its plants and animals, and behave more sustainably. There is evidence from global studies that such interventions can be effective (Moss, Jensen & Gusset, 2015). Such activities can even deliver long-term environmental learning impacts (Jensen, Moss & Gusset, 2017). However, evidence also exists that general learning about the environment through settings such as zoos and aquariums may have only a relatively limited effect on pro-conservation behavior (Moss, Jensen & Gusset, 2017b).

9.2.5 *Public Relations and Marketing Professionals*

Over the past decade, sustainability has also become a term used in corporate communication and marketing to communicate about the advantages of products, processes, or services (Allen, 2016). Marketeers build on the principle that environmental sustainability is widely considered a thing to strive for in consumers' eyes. If the product is sustainable, consumers may think the product has added value and they will, therefore, be more willing to buy it. For instance, since 2018 in the Netherlands a big supermarket chain has integrated messages about its sustainable behavior into its advertising campaign. For example, the campaign speaks about the supermarket using less plastic wrapping for their products, so less use of resources, less production of plastic waste, which equals more sustainability. The campaign also points out that by buying more locally grown products, the level of food transportation needed is reduced.

In addition, the concept of social marketing — that is, marketing for actions or policies that are socially accepted as good — has gained favor within conservation organizations. Often, such efforts seek to establish a pro-environment social norm, that is, the sense that the whole community is supportive of a given pro-environmental action. For example, an international social marketing-focused conservation group called Rare specializes in pride campaigns. It explains its approach on the homepage of their website as follows: 'Rare's signature Pride campaigns inspire pride around unique natural assets and create a clear path for local change. Rare disseminates messages to evoke the desired behavior just like the private sector has done for years' (Rare, 2019). The campaigns often involve a focus on a particular local animal species that Rare seeks to build enthusiasm for as a vehicle to spark pro-environmental change (see Figure 9.1).

There is reason to believe that campaigns such as those of rare can be useful activities because such targeted environmental communication campaigns have been shown to be effective at a global scale (Moss, Jensen & Gusset, 2017a). However, it is always worth bearing in mind that there can be a sizeable gap between norms, awareness, and attitudes on the one hand and behavior and structural change on the other hand (e.g. Moss, Jensen & Gusset, 2017b).

Figure 9.1: A pride campaign parade in the streets of Bitaco, Colombia (photo credit: Rare).

9.2.6 *Policy-makers*

Environmental policy is a rapidly evolving domain around the world, with both national and global efforts underway to enhance environmental sustainability through government interventions. One example where new approaches to environmental governance and policy-making are being implemented is in policy that aims at mitigating the consequences of climate change. Climate change is a typical scientific controversy with contested knowledge and both economic and political stakes involved. The uncertainties pertain not only to technical data or research results but also to ethical and moral aspects, such as, how the interests and wealth of different countries, different cultures, next generations, flora and fauna, should be weighed up (see also Chapter 4)? By keeping all those aspects in mind and involving many different groups of stakeholders, informed by the regular reports of the IPCC, world leaders are trying to make robust global environmental policy.

9.3 Understanding Audiences for Environmental Communication

Each of the categories of practitioners above are seeking to engage audiences with environmental issues, views, values, challenges, and solutions. Important target groups for sustainable or green behavior include, for example, industry, which may be compelled by public pressure or government action to clean up their processes. After all, renewed reasons to start talking about protecting the environment were unwanted consequences of industrialization, such as pollution in the form of acid rain or radioactive contamination of the local environment due to nuclear accidents. A second important target group is society and citizens. They have been the focus of many public campaigns trying to empower them to reduce and reuse waste and to participate in energy-saving schemes.

A particular challenge for motivating individuals to start acting pro-environmentally is that a single individual will not be able to make a change. Pro-environmental measures will only be effective if a large majority of society participates, and this makes it a lot harder to meet the aim of more sustainable behavior. Changes will have to be made for the greater good and not for direct personal gain, such as is the case in health communication (see Chapter 8).

An interesting development regarding audience is that, increasingly, individuals turn from receivers into senders, because they become involved in public debates around certain local environmental issues or because they become authors themselves via social media on environmental issues that they have specialist lay expertise about.

Therefore, when talking about the environment with different target audiences and trying to support pro-environmental behavior, more information than the bare scientific and technological ideas, facts, or data are important. Beliefs and emotions are essential too, both affecting the (communication of) science content and how people view nature and the world around them. It is helpful to understand the different views, values, and norms people hold, and how these perspectives can influence the understanding of or the discussion about the environment. Box 9.1 illustrates the presence and impact of people having different views on the contentious environmental issue of whether to feed wild animals during harsh winters.

Box 9.1: Environmental communication and managing nature

A discussion about the way nature should be managed in an important nature park in the center of the Netherlands is an example of how different views of nature and values can underpin perceptions and spark highly emotional debates about environmental issues. This Dutch nature area developed over the past decades into a relatively 'wild' park where humans are kept out. Large grazers such as deer, wild horse, and cattle breeds are keeping the marshy, flat land clear and, in this way, contribute to the development of a habitat suitable for many birds and land and water animals.

In 2017–2018, the cattle, deer, and horses faced a shortage of food, leading to death by starvation. The Dutch National Forest Management agency that carries out nature management in the park decided to let nature take its course, and withheld intervention. This decision can be seen as part of the value system that views 'nature as simply needing to be left alone', with the park being as close to wilderness and as natural as is possible in the managed landscape of the Netherlands.

At the same time, however, action groups, mainly consisting of cattle farmers, horse breeders, and animal welfare and protection groups, became very vocal in demanding that the animals should get extra food during bad winters. They also acted upon their demands, staging demonstrations, and some of them entered the restricted areas of the nature area to give the animals extra hay. Their stance and actions can be seen as reflecting the worldview that nature is a resource for which humans are responsible.

This example highlights the fact that people are motivated by different general worldviews, views of nature and values, which then underpin their specific stance on particular issues. Therefore, it is important to analyze and understand such perceptions to inform the development of communication products. Moreover, if changing hearts and minds is the goal, values and emotions need to be integrated into the design of effective environmental communication interventions.

9.4 Interventions Stimulating Pro-environment Action

Within environmental communication, a key aim is to promote and support pro-environmental behavior or wider structural changes that are beneficial and supportive for a good, diverse, and healthy environment. One way to try and achieve this aim is to take a systemic or social change approach (Wagoner, Jensen & Oldmeadow, 2012). Those employing a social change approach want to transform the society and economy to become more environmentally sustainable. An example of this would be the 2015 European Union (EU) directive on plastic bags. This directive requires national EU governments to either ban shops from offering free lightweight plastic bags to customers or devise other methods to curb plastic bag consumption. EU governments have taken action with a dramatic decline in the use of plastic bags as a result.

Within a systemic or social change approach, a range of interventions can be used to develop the desired pro-environmental change or effect. This range of interventions includes legislation and regulation, pro-environmental incentives, and public communication campaigns.

9.4.1 *Legislation and Regulation*

Worldwide, many public policy initiatives intend to benefit the environment. For example, many countries promote the use of electric cars by offering tax benefits. Regulations often require cars and industrial vehicles to be checked to ensure exhaust fumes are clean enough to be released into the environment. Some countries subsidize the purchase of renewable energy supply by citizens. And industry pollution is often closely regulated to protect the environment. If communicated and enforced effectively, such policies can deliver permanent improvements in the environmental sustainability of private citizens and organizations large and small.

9.4.2 *Incentives for Environmental Sustainability*

Nudging people toward more pro-environmental behavior through small changes in physical structures or materials can be an important strategy. For example, coffee shops can encourage people by giving them a choice to (re)use their own cup if they are getting coffee from a vending machine. Another example is facilitating people to take the bike instead of the car for short-distance travel. Some employers in countries such as Denmark offer a work benefit allowing employees to pay for a new bicycle or an annual public transport pass with an interest-free loan that is paid back automatically out of their paychecks.

9.4.3 *Public Communication Campaigns*

Public campaigns are a common approach within environmental communication, often aimed at increasing awareness or changing norms to be more sustainable. For example, wildlife conservation campaigns may seek to boost positive attitudes about a local, endangered animal species to build support for its conservation. Or, a campaign among university students may seek to encourage reduced energy use by turning off lights and appliances when not in use in lecture halls or labs.

In the last few decades of the 20th century, public campaigns were often mass media based and strongly sender driven. These campaigns targeted the individual with the aim of creating more awareness and stronger pro-environmental sentiments. For instance, TV ads showed the impact of acid rain with footage of dwindling pine forests in Eastern Europe. This way it was thought people would be motivated to commute less by car.

Around the turn of the 21st century, research showed that the impact of such sender-driven, mass media public campaigns in environmental communication was relatively small. Quite a few environment-related problems had increased in size, impact, and complexity over the years with many different stakeholders involved. To tackle such complex issues, new approaches were needed and developed. In addition to the more sender-driven public campaigns, new formats emerged with a focus on involving and engaging their target audiences, turning them into participants rather than approaching them as mere receivers of a message.

One powerful example of an environmental campaign in which the academic research into the problem was combined with public funding and the input and help from local people is the Loess plateau campaign (Wei *et al.*, 2006). At the end of the 20th century, this plateau in China's northern region was a barren space due to soil erosion and unsustainable agricultural methods. Within two decades, the Chinese government, together with the local farmers, turned this dry and harsh area into an oasis. Initial input by local farmers was instrumental in figuring out what had happened to the region to make it so unfertile and uninhabitable. On the basis of that information, an elaborate landscape restoration plan was set up, in which again the local people were crucial partners. In measuring the initial attitude of the farmers to the project, researchers discovered that one element was crucial: providing them with (new) means to feed their families. Thus, the plan was set up to pay the local people for working on the restoration project. The local people labored on the allotments, constructed terraces for agricultural practices, planted trees, and built soil traps and dams. In addition, during the following 10 years, their work on maintaining all these structures remained subsidized.

The Loess plateau example is a successful one because it takes into account the needs of the people who are most essential to developing the pro-environmental change. Because environmental problems are so big and global, it is easy for people to think, 'if I stop driving my car or become vegetarian it is not going to make a difference'. Therefore, providing a clear incentive — whether social or economic — in addition to good communication, can be very helpful to prompt and maintain positive action by the target audience.

9.5 Models to Guide Interventions

Indeed, environmental communication interventions often aim to foster some kind of change. In addition to a pro-environmental legislation and regulation framework and establishing incentives, a systematic approach to every phase of communication intervention development, implementation, and evaluation is necessary to increase the probability that efforts, such as a public campaign, will be successful.

A well-researched and much used systematic approach that is used in health communication is Intervention Mapping (Bartholomew-Eldredge *et al.*, 2016; see also Chapter 8). As health communication interventions, just like environmental communication interventions, often aim to create (behavioral) change, Intervention Mapping can be a helpful model for environmental communication.

In health communication, Intervention Mapping is a planning process for the systematic development of health promotion interventions, in which an individual's larger social context is taken into account. This approach works from problem identification to problem solving by repeatedly going through the first four of six steps as follows:

1. initiating a problem analysis, also called needs assessment;
2. devising program aim(s);
3. developing methods, strategies, and activities;
4. establishing coherent program development;
5. preparing an implementation and adoption plan; and
6. creating an evaluation plan.

Kok *et al.* (2011) applied Intervention Mapping to the domain of energy conservation, focusing on two existing, (relatively) successful energy conservation intervention studies. The first one aimed to encourage household energy conservation, for example, shorter showering or turning thermostats low in empty rooms. The second intervention study sought to encourage mail van drivers to implement a new, fuel-saving driving style.

Using Intervention Mapping helped planners in both intervention studies to identify the determinants of behaviors and related context (e.g. barriers, needs) related to the energy conservation problems. Important personal factors identified in the household energy case include knowledge of energy conservation, awareness of the need for it, the person's attitude toward energy conservation, having goals for energy conservation, quality of experience trying out certain energy conservation methods, self-perception about whether a person is capable to carry out the desired behavior, and comfort and price of particular energy conservation methods (Uitdenbogerd *et al.*, 2007). Important contextual, thus non-personal more external, factors included the ability to choose out of a selection of possible energy conservation methods, income of the person, the type of housing lived in, building-related measures, and technical provisions.

In the intervention study focusing on encouraging fuel-saving driving styles, key personal factors included intention, attitude and beliefs, and personal norms of the person toward fuel-saving driving styles, perceived social norms, in other words, what important others think of fuel-saving driving styles, self-efficacy, and past behavior, for instance, a person's own idea about being capable of carrying out a fuel-saving driving style (Lo, Peters & Kok, 2012). Contextual factors, the non-personal more external factors in the fuel-saving driving study case, included trust in the bosses and management of the company, the presence of pro-environmental policies implemented in the company, bosses implementing a pro-environmental management style, and making it easy for people follow a pro-environmental behavior in other ways, for instance, by placing easily accessible recycling bins on the work floor.

Based on the personal and contextual factors that were identified, a suitable theoretically grounded intervention could be selected. For instance, in the case of the mail van drivers, the theoretically informed approach of displaying the desired behavior to encourage learning through observation was used (Bartholomew *et al.*, 2006; Bartholomew-Eldredge *et al.*, 2016). A fellow driver would demonstrate the new driving style, showing a reduction in fuel use if a car was driven over the same distance, and within the same time, while applying the fuel-saving driving style. Developing such evaluation evidence is clearly valuable for refining an environmental communication intervention to ensure it is as effective as possible (Jensen, 2015).

Kok *et al.* (2011, p. 5285) concluded that: 'Although [Intervention Mapping] is a complex and time-consuming process, the benefits of its consistent application may outweigh its costs by ensuring more effectiveness and efficient learning through its evaluation processes.' It is important to use models such as Intervention Mapping to establish a systematic approach to develop environmental communication interventions.

9.6 Developing Theoretically Informed Approaches

Environmental communication often focuses on overcoming personal, social, and economic barriers to facilitate desirable pro-environmental behavior. In developing interventions such as public campaigns, evidence-based models and theories often provide valuable insights. That is why widely used socio-psychological theories that explain how behavior change develops can be very useful for environmental communicators. However, over the past decades, research has shown that behavior is very hard to predict, let alone change, and causal relationships between communication and behavior can be challenging to demonstrate.

An influential theory in the field of environmental communication (as it is in Chapter 8) is the theory of planned behavior, also developed as the theory of reasoned behavior (Fishbein & Ajzen, 1975; Ajzen, 1991). According to the perspective underpinning these theories, behavioral intention is the most important predictor of behavior that is based on active decision-making. For instance, tourists can decide no longer to fly to a holiday destination in order to reduce their carbon footprint. If they have this behavioral intention, it is more likely that they will behave accordingly, and next time consider taking a train (or other more sustainable transport option) instead of a plane. Underlying determinants that explain the formation of behavioral intentions are attitudes toward the desired behavior, social norms, thus what important others think how one should behave, and self-efficacy (i.e. the feeling of personal empowerment that one is capable of carrying out the desired effect and that it will have an effect or make a difference).

This theoretical perspective can be illustrated by the example of trying to motivate 14- to-17-year-old secondary school pupils to use less plastic by not drinking with plastic straws and reducing littering. Examples of attitudes toward the behavior in this case could be: 'Yes, the plastic problem is really big and something should be done about it, but I really enjoy drinking soft drinks with a straw', or 'I like the pre-wrapped sandwiches at school better than a home-made sandwich taken to school in a lunch box'.

Social norms represent what people believe are the typical behaviors in their social circle, for example, among family members and close friends. To stick with the example, a relevant social norm might be: 'Most of my friends buy pre-packaged, take away lunches at school, which always have a plastic wrapper around them'. Social norms also pertain to the subsequent approval or disapproval by others that would follow their actions. Factors such as perceived social support or social pressure are also part of social norms. For example, a statement pertaining to social norms could be: 'My best friend always separates the plastics from household waste, and keeps on lecturing me to do the same'. Young people are often more influenced by social norms among their peer group than by those of their family members, for instance, their parents.

Self-efficacy involves an assessment of whether one is able to successfully carry out the desired behavior. Aspects such as perceived difficulty and self-confidence play a role here. A statement exemplifying self-efficacy would be: 'There is nothing I can do about the problem of plastic in the environment', or 'I am confident I can make a difference by putting plastic in the designated recycling bin'.

According to the theory of planned behavior, relevant skills help people to act on their behavioral intentions, for instance, organizational skills, while structural barriers hinder follow-through from intention to reality. For instance, a barrier such as the city council not providing separate collection of plastic waste for recycling could make it hard for people to work toward reducing plastic waste.

Fishbein and Azjen's theory of reasoned behavior works well when people think about their intentions and make a conscious decision before taking action. But that is often not the case in everyday life. A lot of things people do are based on habit, not on thought–through reasoning. This habitual behavior is determined largely by people's unarticulated perceptions, values, and emotions (see also Section 9.3). That is why, in the past decades many researchers have adapted this influential theory so it may fit better particular situations where unarticulated perceptions, values, and emotions play a role, and in which overcoming personal, social, and economic barriers to facilitate desirable behavior is important. Think, for instance, about fostering a healthy or more sustainable lifestyle (see also Chapter 8).

Sometimes, a gap between people's attitudes about something like climate change and their behavior (e.g. taking flights for holidays) exists. For instance, a person just loves flying and going on holidays to exotic destinations but at the same time believes people should reduce their carbon footprint. Such a tourist ends up in a state of cognitive dissonance (Festinger, 1957). Research on cognitive dissonance shows that people in such a dilemma tend to change their attitude rather than their behavior, for instance, by thinking: 'Carbon footprint is not the most important issue, actually biodiversity and poverty are the real issues'.

9.7 Confronting Grand Environmental Challenges

In the past decade, in many developed countries headway has been made toward more sustainable living and working. Infrastructure for recycling has made it feasible for citizens in many parts of the world, supported by increasingly pro-sustainability social norms, to reduce their negative impact on the environment. However, many of the really challenging issues have not yet been dealt with, such as the clash between the overriding desire for economic growth through production and sale of new consumer goods and the pro-sustainability imperative to reduce consumption by repairing and reusing existing goods.

Addressing the issue of climate change is one of the defining challenges of the 21st century (and beyond), but it has been very difficult to get political traction for the necessary global action. Surveys show that climate change risks are generally not one of the main worries of the general public (Steentjes *et al.*, 2017). The lack of public concern about climate change is partly due to differences in risk perception of the general public and the scientific experts (see Chapter 7). So successfully appealing to individual citizens to take action on a purely voluntary basis, for example, by no longer taking a plane when going on holiday, or by eating less beef, or by buying more sustainable (but more expensive) consumer goods, or doing without some consumer products entirely, is very difficult.

Expert organizations, such as the IPCC, clearly indicate that climate change is real and desperately needs to be curbed to limit devastating global consequences. Yet, people experience climate change as a far-off and diffuse problem. Moreover, those with an economic interest in avoiding serious legislative and regulatory action on climate change have cultivated the notion that there is a lack of scientific consensus about the problem or solutions. Moreover, because action has to be taken on a worldwide level, individual countries are reluctant to take responsibility for this shared problem.

With such complex and evolving scientific issues, it is good to discuss the science, perspectives, and worries openly, address the uncertainties, and involve and engage as many stakeholders as possible, including citizens. Moreover, it is important to take into account audience knowledge, perceptions, values, and emotions in environmental communication in order to engage effectively (see also Section 9.3).

In this context, it is helpful to draw on recent research on climate change communication (e.g. Nisbet *et al.*, 2018). One of the promising findings in this domain is the potential impact of framing issues such as climate change in new ways (see also Chapter 6), such as the frame of economic opportunity (Nisbet & Scheufele, 2009), public health, or national security (Myers *et al.*, 2012).

Another way of making communication more personal and relevant for the public is through storytelling. Research shows, even though with mixed results, that stories can influence beliefs, attitudes, intentions, and behaviors. The more these stories are tailored to people's worldviews, emotions, and beliefs, the higher the chance of success (Braddock & Dillard, 2016).

9.8 Conclusion

In recent years, the context for environmental communication has changed dramatically. In terms of content, the UN SDGs are on the rise, and the global challenge of climate change is increasingly seen as the most important single environmental issue to address. In terms of platforms for communication, the growth and diversification of social media is a key development, along with the changing nature of science and environmental journalism.

In terms of how to communicate effectively, perspectives such as the theories of planned or reasoned behavior, models such as Intervention Mapping, and specific techniques such as evidence-based message framing offer valuable insights that can inform practice.

Just as with other forms of communication, it is essential to analyze audience values and needs in order to develop effective interventions for engaging people with environmental issues. Building on ongoing research and evaluation, ever-more effective environmental communication approaches can be developed to contribute to some of the world's most urgent and far-reaching challenges.

References

Ajzen, I. (1991). The theory of planned behavior. *Organizational Behavior and Human Decision Processes, 50*, 79–211.

Allen, M. (2016). *Strategic Communication for Sustainable Organization: Theory and Practice.* London: Springer.

Bartholomew, L. K., Parcel, G. S., Kok, G., & Gottlieb, N. H. (2006). *Planning Health Promotion Programs: Intervention Mapping* (2nd edn.). San Francisco, CA: Jossey-Bass.

Bartholomew-Eldredge, L. K., Markham, C. M., Ruiter, R. A. C., Fernández, M. E., Kok, G., & Parcel, G. S. (Eds.) (2016). *Planning Health Promotion Programmes: An Intervention Mapping Approach* (4th edn.). San Francisco, CA: Jossey-Bass.

Bennett, N. J., Roth, R., Klain, S. C., Chan, K., Christie, P., Clark, D. A., Cullman, G., Curran, D., Durbin, T. J., Epstein, G., Greenberg, A., Nelson, M. P., Sandlos, J., Stedman, R., Teel, T. L., Thomas, R., Veríssimo, D., & Wyborn, C. (2017). Conservation social science: Understanding and integrating human dimensions to improve conservation. *Biological Conservation, 205,* 93–108.

Braddock, K., & Dillard, J. P. (2016). Meta-analytic evidence for the persuasive effect of narratives on beliefs, attitudes, intentions, and behaviors. *Communication Monographs, 83*(4), 446–467.

Bucchi, M., & Trench, B. (2014). Science communication research: themes and challenges. In M. Bucchi & B. Trench (eds.), *Routledge Handbook of Public Communication of Science and Technology.* London: Routledge, pp. 1–14.

Festinger, L. (1957). *A Theory of Cognitive Dissonance.* Stanford, CA: Stanford University Press.

Fishbein, M., & Ajzen, I. (1975). *Belief, Attitude, Intention and Behavior.* New York: Wiley.

Greenpeace. (2018). *Greenpeace International website.* Retrieved December 20, 2018 from http://www.greenpeace.org/international.

Horst, M., Davies S. R., & Irwin, A. (2017). Reframing Science Communication. In U. Felt, R. Fouché, C. Miller, & L. Smith-Doerr (eds.), *The Handbook of Science and Technology Studies.* Cambridge, MA: MIT Press, pp. 881–907.

IECA. (2019). *Environmental Communication: What It Is and Why It Matters.* Retrieved January 27, 2019 from http://theieca.org/resources/environmental-communication-what-it-and-why-it-matters.

Irwin, A., Bucchi, M., Felt, U., Smallman, M., & Yearley, S. (2018). *Re-framing Environmental Communication: Engagement, understanding and action.* A Background Paper for MISTRA. 1–23.

Jensen, E. (2015). Evaluating impact and quality of experience in the 21ˢᵗ century: Using technology to narrow the gap between science communication research and practice. *JCOM: Journal of Science Communication, 14*(3), C05. Last accessed 27 January 2019, Retrieved January 27, 2019 from http://jcom.sissa.it/archive/14/03/JCOM_1403_2015_C01/JCOM_1403_2015_C05.

Jensen, E., Moss, A., & Gusset, M. (2017). Quantifying long-term impact of zoo and aquarium visits on biodiversity-related learning outcomes. *Zoo Biology*, *36*(4), 294–297. doi: 10.1002/zoo.21372.

Kok, G. N., Lo, S. H., Peters, G. Y., & Ruiter, R. A. C. (2011). Changing energy-related behavior: An Intervention Mapping approach. *Energy Policy*, *39*, 5280–5286.

Lo, S. H., Peters, G. J., & Kok, G. (2012). A review of determinants of and interventions for pro-environmental behaviors in organizations. *Journal of Applied Social Psychology*, *42*(12), 2933–2967. doi.org/10.1111/j.1559-1816.2012.00969.x.

Moss, A., Jensen, E., & Gusset, M. (2015). Evaluating the contribution of zoos and aquariums to Aichi Biodiversity Target 1. *Conservation Biology*, *29*(2), 537–544. doi: 10.1111/cobi.12383.

Moss, A., Jensen, E., & Gusset, M. (2017a). Evaluating the impact of a global biodiversity education campaign on zoo and aquarium visitors. *Frontiers in Ecology & the Environment*, *15*(5), 243–247. doi: 10.1002/fee.1493.

Moss, A., Jensen, E., & Gusset, M. (2017b). Probing the link between biodiversity-related knowledge and self-reported pro-conservation behavior in a global survey of zoo visitors. *Conservation Letters*, *10*(1), 33–40. doi: 10.1111/conl.12233.

Myers, T. A., Nisbet, M. C., Maibach E. W., & Leiserowitz, A. A. (2012). A public health frame arouses hopeful emotions about climate change, A Letter. *Climatic Change*, *113*, 1105–1112. doi: 10.1007/s10584-012-0513-6.

Nisbet, M. C., & Scheufele, D. A. (2009). What's next for science communication? Promising directions and lingering distractions. *American Journal of Botany*, *96*(10), 1767–1778. doi.org/10.3732/ajb.0900041.

Nisbet, M. C. (Ed.), Ho, S., Markowitz, E., O'Neill, S., Schafer, M., & Thaker, J. T. (Ass. eds.), (2018). *The Oxford Encyclopedia of Climate Change Communication*. New York: Oxford University Press.

Rare. (2019). *Rare website*. Retrieved January 31, 2019 from https://www.rare.org/pride.

Smallman, M. (2014). Public understanding of science in turbulent times III: Deficit to dialogue, champions to critics. *Public Understanding of Science*, *25*(2), 186–197.

Steentjes, K., Pidgeon, N. F., Poortinga, W., Corner, A. J., Arnold, A., Böhm, G., Mays, C., Poumadère, M., Ruddat, M., Scheer, D., Sonnberger, M., Tvinnereim, E.

(2017). *European Perceptions of Climate Change (EPCC): Topline findings of a survey conducted in four European countries in 2016.* Cardiff: Cardiff University. Retrieved January 27, 2019 from orca.cf.ac.uk/98660/7/EPCC.pdf.

Uitdenbogerd, D., Egmond, C., Jonkers, R., & Kok, G. (2007). Energy-related intervention success factors: a literature review. In *Proceedings of the ECEEE Summer Studies of the European Council for an Energy Efficient Economy.* Retrieved December 15, 2018 from http://www.eceee.org/conference_proceedings/eceee/2007/Panel_9/9.040/

Wagoner, B., Jensen, E., & Oldmeadow, J. (eds.), (2012). *Culture and Social Change: Transforming Society Through the Power of Ideas.* Charlotte, NC: Information Age Publishers.

Wei, J., Zhou, J., Tian, J., He, X., & Tang, K. (2006). Decoupling soil erosion and human activities on the Chinese Loess plateau in the 20th Century. *Catena, 68*(10), 10–15. doi.org/10.1016/j.catena.2006.04.011.

World Commission on Environment and Development. (1987). *Our Common Future.* Oxford: Oxford University Press.

Chapter 10

Research in Science Communication

Anne M. Dijkstra and Craig Cormick

10.1 Introduction

The dynamic and growing interaction between science and technology and society has led to an increased drive to communicate about science, and to undertake research how to do this effectively. Sometimes, research into science communication can even have a life-or-death importance, such as when it helps professionals understand why some people refuse mainstream medicine in favor of alternative therapies. Alternatively, it can explain why some people refuse to vaccinate their children, potentially putting them at risk of life-threatening diseases. In such cases, communicating science and technology is not about trying to ensure economic development or technological diffusion but rather to promote community well-being and individual safety (see also Chapter 8).

In the past few decades, science communication has emerged as an area of research and practice in its own right (e.g. Guenther & Joubert, 2017; Trench & Bucchi, 2015). However, science communication is also viewed differently by many different people working in the field. One of the peculiarities of science communication is that people have come to it from many different fields, and they have brought ways of thinking and acting from those fields, including education, social studies of science, mass communication, psychology, or other fields (see also Mulder, Longnecker & Davis, 2008).

Despite the diversity, two key types of communities work in science communication — researchers and practitioners — and tensions between them can have impact on the research. Science communication practitioners undertake science communication activities as their core task, while science communication researchers or scholars conduct research into science communication. Traditionally, both groups have not interacted as well as they could, even though such interaction would benefit both. Efforts to change that are, for example, initiated by the Public Communication of Science and Technology network (pcst.co) which organizes its biennial conferences for both scholars and practitioners.

For scholars, conducting research defines their work, while practitioners see undertaking science communication activities as the most important thing. Yet, science communication activities could improve significantly if these were more often based on research being undertaken. Nisbet & Scheufele (2009), for example, argue that research findings can contribute to understanding how modern societies make sense of science and technology, which would be of significant benefit for practitioners and researchers both. They state that the basic premise for doing research should be that 'any science communication effort needs to be based on a systematic empirical understanding of an intended audience's existing values, knowledge and attitudes, their interpersonal and social contexts, and their preferred media sources and communication channels' (p. 1).

Of course, the fields of science communication practice and research have evolved over the years, going from a more simplistic focus on raising awareness and understanding to address a deficit of knowledge through to more two-way engagements built around working with people's differing values, receptiveness, and needs for information (Bauer, 2009; Bauer, Allum & Miller, 2007). And as research has evolved in complexity, building on what has come before it, practice eventually evolves with it.

This chapter presents how research insights in science communication processes and products can help both researchers as well as practitioners to communicate about science more effectively. It will not be a detailed how-to-do-science-communication-research chapter, as that will be beyond the scope of this book. It will just describe important steps in the research process, main research methods, and research ethics issues and will look at some research questions that still need to be addressed.

10.2 The Research Process

Regardless of the objective of conducting science communication research, designing and carrying it out will follow roughly the steps as described below (based on, for example, Jensen & Laurie, 2016; Wilkinson & Weitkamp, 2016). It is good to keep in mind that, often, doing science communication research — as any other research process — is an iterative process where it is possible to go back and forth multiple times.

a. Identify the research topic and research aims
The first stage in the research cycle is to identify the research topic and clearly articulate the research project's aims. An example could be what science and technology messages people take away from news reports as opposed to what messages they take away from science documentaries. These could then be analyzed to see if the messages were similar or at odds with each other. Often, the problem at hand needs to be explored in more detail through looking at existing research or literature so the topic for research can be more precisely identified. This can also help to provide a rationale and framework for the research work. An example might be concentrating on how climate change is portrayed in news reports or documentaries and comparing coverage to the information of the original scientific reports. An analysis of existing literature might show that some work had been done on this for television, but not for newspapers, so the research could then not just compare what the original scientific data showed with newspaper reports, but analyze how this differed from television coverage.

b. Formulate a research question
In a next step, the literature is further explored and a main research question is formulated. Possibly, subquestions will be added. A research question, ideally, can be based on a gap in the literature, can be answered by the proposed research, is more than a description, and is clear and concisely formulated. An example could be whether people's values differ based on how they consider life sciences such as biotechnology and non-life sciences such as nanotechnology, or what messages they take from different news reports or documentaries about such topics.

c. Choose a research method

In the next step, the method (or methods) which is most suitable to answer the research question is decided upon. A choice will be made between qualitative or quantitative research methods, or a combination of methods — a mixed methodology (see also Section 10.3). The most common methods used are surveys, interviews, focus groups and case studies, but there are other less conventional methods that are also used, such as having members of the public undertake drawings, graffiti walls, or diaries, to gain an understanding of their thinking beyond standard methods (Wilkinson & Weitkamp, 2016) (see also Section 10.4.4 and Box 10.1 for more information about research methods). In practice, more often than not, the chosen method will also depend on the limited available budget and time, and sometimes, the experiences of the researchers who will conduct the research. If the research needs to be undertaken with members of the public and also when it involves, for example, a media analysis, then decisions will be taken on the sample (e.g. size). When people are involved, it is also important to consider how to arrange ethical approval and ask for informed consent and set up the procedures for this. In addition, researchers have to determine how to anonymize the data to remove identities and how to store this information in a safe place once the data are collected (see Box 10.1).

d. Collect data

After working out the practicalities regarding the method or methods (such as how to find respondents or research participants), data can be collected. This can involve recruitment via a specialist company or by seeking volunteers, or by recruiting people at public events. Data collection can be done via online methods, direct engagement, or a number of other means. The choice of what method to use can often be based on comparing with an earlier study as well as expertise and experience of the researcher and other available resources.

e. Analyze data

In the next step, the data will be analyzed. Before that can be done, the data set will have to be prepared, for example, by transcribing and coding the data, or preparing the data set for statistical analysis. This might also include transcribing recordings or writing up notes in a way that can be more easily

Box 10.1: Research methods in science communication research

Key methodologies for conducting research include, first, the choice for qualitative or quantitative data collection, or a combination, and, thereupon, the method. The main methods are surveys, interviews, focus group discussions, and case studies, which together represent the most used in social scientific research.

Survey questionnaire — can also be conducted online
Surveys are a set of questions which will be answered by the research participants. The format allows for large groups to be included and consists often of mainly closed questions and a few open-ended questions. Based on the literature and researchers' insights, questions are included or excluded. A set of questions related to one topic often is more robust than single items. Surveys can be filled in via face-to-face interviews or, nowadays, online and they get the best uptake if they take a maximum of 10 minutes to complete.

Interview
Interviews are conducted in a one-to-one setting, with research participants answering semi-structured, or open questions. Interviews can provide in-depth information. Under normal circumstances, between 15 and 20 different interviews should gain enough different perspectives around a research topic, though in large studies, hundreds of interviews might be conducted.

Focus group discussion
A group of 6–8 participants will talk about a research topic and answer semi-structured or open questions for about 2 hours usually, under the guidance of a moderator. Focus groups enable in-depth insight on the various aspects around a topic and allow participants to respond to each other's points, which single interviews cannot achieve. This method is useful for complex topics or to obtain a variety of views and opinions. Focus groups rely on a skilled moderator who can guide the discussion in such a way that it is not dominated by one or two participants (see also Chapter 4 on the different roles for a dialogue moderator).

Case study
In a case study, a specific topic will be explored in more detail often using a combination of document analysis, interviews, and possibly other methods. Conducted in a systematic way, a detailed analysis provides in-depth

(*Continued*)

Box 10.1: **(Continued)**

information of the case. Sometimes, multiple cases are compared with each other.

Other research methods

Research methods such as a systematic literature review and document analysis do not use research participants directly, but study documents. In addition, plenty of other methods exist which include research participants but are less frequently used, such as ethnographic observations (observing a situation) and participatory evaluation (evaluate while also participating in the meeting or event). Also, in recent years methods from the field of Science and Technology Studies have been included in science communication research. For instance, research methods such as constructive technology assessment, where a technology is assessed in its social context or midstream modulation (Schuurbiers & Fisher, 2009), include multiple interviews with researchers and can be used to answer research questions that seek to answer the role of technology in society. Less common in science communication research are social/ psychological experimental research studies which are conducted in research laboratories.

Systematic literature review

All studies will contain a short literature review, however, in a systematic literature review, this is taken a step further. The academic literature on a topic is collected and often a sample is analyzed in a more systematic way. This method can provide an extensive overview of the main aspects around a topic. Petticrew & Roberts (2006) describe steps for a systematic literature review in more detail.

Document analysis

Documents other than published journals and books that relate to a topic can be collected and analyzed. Often referred to as 'gray literature', it can include research studies, reports, newsletters, and news reports. A document analysis can provide important additional information on a research topic. It often serves as a preparation for other research methods such as a case study.

Less conventional methods

Wilkinson & Weitkamp (2016) mention less conventional methods which may be of interest. It is also possible to analyze comments in guest books,

---

Box 10.1: (*Continued*)

drawings, charts, mind maps, and diagrams that research participants produce, for example, when visiting a science museum or a science festival. Diaries can give more personal opinions of topics. In addition, graffiti walls, images, photographs, and videos all used in combination with short interviews can answer research questions about how one feels about something beyond what they are able to articulate in an interview.

Citizen science
Finally, in citizen science, scientists use help from citizens to collect, analyze, or interpret findings. This particular method of research is on the rise globally. Chapter 5, Informal Science Education, presents an overview of the diversity of possibilities and potential drawbacks of research using or based on citizen science.

Sources: Based on studies such as Wilkinson & Weitkamp (2016); Chapters 10 and 11; Jensen & Laurie (2016).

---

analyzed. Programs for both qualitative as well as quantitative analysis can be used to assist the process, although analysis can often be done without advanced programs.

f. Report the outcomes

Finally, the outcomes or findings of the research will need to be written up, which may lead to further reflection upon them. It is then time to draw conclusions and discuss these in relation to the main aims formulated in the beginning of the study. A research report or article will normally contain the following into chapters or sections: introduction, literature review or theoretical framework, methods, results, discussion, and conclusion.

Many reasons exist for undertaking science communication research. It can be driven by the objectives of individual researchers, institutions, or funding agencies, which can impact what research is conducted. For example, following funding stream changes across the EU and OECD, research that had previously been focused on better understanding social impacts and choices of genetic technologies moved toward better understanding social impacts and choices around nanotechnologies.

Also, individual and institutional factors can govern how research is conducted and what type of data collection is used. For instance, some individual researchers have a strong preference for the analysis of large sets of data and prefer quantitative research. Quantitative research is based on collecting large quantities of data, for instance, via online polling. Others prefer qualitative data, based on collecting smaller data sets that are more detailed, such as through focus group discussions and individual interviews.

Sometimes, the end use of the data can determine what type of research is preferred. Governments, for example, who regularly use voter polling to determine public attitudes to issues, tend to prefer quantitative data over qualitative data, while commercial companies testing products on consumers often prefer qualitative data that is more likely to explain how a person feels about something.

10.3 Methodological Pluralism and Multidisciplinary and Interdisciplinary Collaboration

In the past, science communication as a field had many studies which restricted their research methods to either quantitative methods, where large amounts of data were gathered and analyzed, or qualitative methods based on collecting information from small-scale samples, like a few in-depth interviews with a number of experts on an issue (Dijkstra, 2008). Various authors argued that, quantitative surveys, for example, isolate respondents from their social context (e.g. Brewer & Hunter, 2006; Von Grote & Dierkes, 2000; Wynne, 1995). Yet, at the same time, qualitative research methods of social scientific research cannot catch all of the complexities of a situation as well. Brewer & Hunter (2006), consequently, argued that each method has its own weaknesses and strengths and no method is without bias.

It should be understood that all data, irrespective of whether they are qualitative, quantitative, or mixed methodology, are only ever indicative rather than definitive, and it is often important to acknowledge this through error ratings or comparisons with other studies. Proving the validity of a study is important, which means demonstrating that the research measures what it intends to measure. This is often done via statistical analysis of the sample used and demonstrating how well it reflects the broader society.

In line with these thoughts, Von Grote & Dierkes (2000), among others, reasoned that both qualitative research methods as well as quantitative research methods in science communication research are crucial in acquiring better theoretical and practical understanding of science communication concepts and efforts, and they should be used in conjunction with each other (see also Sturgis & Allum, 2004). Both can provide answers and make it possible to acquire enhanced theoretical understanding of processes and outcomes in science communication (e.g. Egan *et al.*, 1995; Greene, Benjamin & Goodyear, 2001). In the end, it is important that the methods best suit the research problem at hand and this can be decided upon by using questions such as whether a very broad understanding is needed or a narrower but deeper understanding is needed.

Due to the nature of the field, and different people's backgrounds, more interdisciplinary (between disciplines) and multidisciplinary (across disciplines) collaboration is possible. For scholars, this is sometimes challenging however, as their articles from such collaboration efforts sometimes are more difficult to publish when they fall outside the scope of a particular discipline. An example of this might be an economic analysis of the types of science communication used by some members of the public. It might be viewed as too economically based for a science communication journal and not economic enough for an economics journal. However, for particular topics where science is contested and debated in the public domain, more insights and better understandings can come from such collaborations.

10.4 Research Ethics

In all research, when dealing with people, it is necessary to explain the steps that are being undertaken, and in doing so treat the research participants with care, sensitivity, and respect. In other words, ensure that no harm is done and risks are minimized (see also Wilkinson & Weitkamp, 2016). This might sound over-cautious, but any one taking part in a study will be affected in some way by that study, whether it is as simple as raising their engagement with a topic or more challenging when it might seek to elicit information they are not comfortable in providing. Ethical issues are by nature complex and multidimensional, so not easy to deal with. They should be part of a research project from start to end, as is, for example, stated on the website of the European Union (https://ec.europa.eu/programmes/horizon2020/en/h2020-section/ethics). Recently, discussions around research ethics and how to deal with ethical issues have received much attention globally as responsible conduct in research and innovation has become a key principle of research projects.

10.4.1 *How to Deal with Ethics?*

Several levels of ethics relate to research. The first is getting ethics approval for research to be conducted. Universities have ethics committees that will usually assess the design of a research project according to the research ethics and the procedural principles in place and will have to approve the study at hand. In the reporting phase, these steps regarding taking care of ethics in the study will be described in the methods section.

A second level of ethics is considering the ethical aspects in practical situations, as it is a part of communicating with participants about the research process (Wilkinson & Weitkamp, 2016). Considering several procedural principles, such as those described in what follows, will help ensure the research is addressing the key ethical issues it might need to cover. For further reference and more detail, several methodology books are available online or in print (e.g. Wilkinson & Weitkamp, 2016).

10.4.2 *Informed Consent and Voluntary Participation*

Informed consent is a way to ensure that researchers provide the research participants with all information which is necessary for them to decide whether they wish to voluntarily participate in the study. This information concerns a summary of the key aspects of the study, for example, who is asked to participate, how the respondents are expected to contribute, how they can withdraw at any time during the study for any reason, and the type of questions that will be asked. It should also convey that opinions expressed will not be associated with them personally, and how anonymity and confidentiality is dealt with. For European countries, for instance, they must conform to the General Data Protection Regulation (GDPR 2016/679) governing data protection and privacy. To ensure that everyone receives the same information, research participants can, for example, be provided with an information sheet and, when applicable, they can be asked to sign an informed consent form.

10.4.3 *Anonymity and Confidentiality*

Anonymity means that the research participant cannot be identified from the reporting while confidentiality refers to the people who might have access to the data. Research participants have the right to know whether data can be traced back to them, as well as to know who is entitled to use or read the data. That is why researchers will anonymize data. Anonymity can be reached by, for example, using numbers or letters or fictional names that represent the research participants. It is also necessary to ensure that research participants cannot be identified by any additional information such as the organization they represent. In those cases where anonymization will not be possible, consent to publish names should be asked beforehand. In addition, how to store and transport the collected data in a safe way needs to be considered and decided upon, for example, by ensuring that the data are password protected at all times.

10.4.4 *Transparency of Sponsorship*

One aspect which also concerns ethical dimensions of research is that of funding. Sometimes, funding bodies may want to influence the research. For research participants it should, therefore, be clear who funds the study or whether the study is a self-funded project. Providing such information in the consent form helps the research participants to consider possible ideological or political objections. This information should also be disclosed at the end of a research article or in a research report.

10.4.5 *Increasing Awareness of and Attention for the Importance of Ethical Conduct*

In many countries and institutions, ethical codes of conduct exist which can be used as the basis for considering ethical dimensions of a study, such as the above-mentioned General Data Protection Regulation that took effect in the European Union from 2018. This regulation provides new and stricter rules than before regarding research ethics for the EU countries, for example, regarding the need to ensure anonymous participation in research studies.

Research ethics is also one of the key themes for responsible innovation, and, increasingly, universities are not only training researchers, among them social scientists, but also scientists and engineers are included, to make them more aware of ethical issues related to conducting research in a broader sense. For example, at the University of Twente in the Netherlands, as in most universities nowadays, a scientific integrity training program has been developed for all PhD students (sciences and social sciences) to increase awareness of ethical behavior and increase knowledge of how to deal with ethical concerns. Also, other staff have been informed about steps to be taken before, during, and after research to ensure responsible research conduct (see Figure 10.1).

10.5 The Research and Practice Divide

An issue that impacts research is that the community of science communication practitioners and the community of science communication researchers don't often work together, nor even for each other's best outcomes (National Academies of Sciences, Engineering, and Medicine, NAS, 2017). This is happening despite a growing number of people who are science communication practitioners, researchers, or scientists, who do research and practice in this field. They may draw on sources as described in Box 10.2.

PERSONAL DATA
RESEARCH PROTOCOL
RESPONSIBLE USE OF PERSONAL DATA & GENERAL DATA PROTECTION REGULATION (GDPR)

BEFORE RESEARCH	Identify Personal Data	Identify Risks	Address in Data Management Plan	Communicate	Register
	Check use of personal data, if so be aware of its lawfulness	Consider risks in using online tools, storage, access	Include details on working with personal data	Be transparent about your processing	Notify Data Protection Officer about use of personal data
	See GDPR at utwente.nl/privacy	If neccesary, use DPIA-Tool	Website: itwente.nl/ researchsupport	utwente.nl/privacy-statement	Website: webapps. utwent.nl/wbpregister

DURING RESEARCH	Implement Rights Data Subject	Get Informed Consent	Anonymise	HOW TO WORK SAFELY	
	Implement procedures to ensure data subjects' rights	Give clear information in consent form. Store informed consents securely	Anonymise personal data as soon as possible	**Storage** Store securely on M:/P: Disk, Sharepoint, SURFDrive	**Surveys** Use safe survey tools
	See utwente.nl/privacy	Use UT Informed Consent Model	See utwente.nl/privacy	**Clear Desk** Don't leave personal data on printer or desk	**Encrypt** Encrypt sensitive data using Bitlocker/Filevault

AFTER RESEARCH	Publish or Present	Archive Data	Be FAIR		
	Publish or present only anonymised or with consent	Minimise archiving of personal data, archive securely	Make anonymised data Findable, Accessible, Interoperable and Reusable	**Transport** Encrypt files in transport	**Report** Report data breaches at cert@utwente.nl
	See UT Informed Consent Model	In line with faculty or group policy	Archive in DANS or 4TU Research Data		

CONTACT: Privacy Contact Person or Data Protection Officer (fg@utwente.nl)
MORE INFORMATION: www.utwente.nl/privacy

Figure 10.1: Personal data research protocol at the University of Twente.

10.5.1 *Differences in Cultures of Learning*

In addition to the described divide in Box 10.2, research directions can also be impacted by other priorities, such as culture, as Guenther & Joubert (2017, p. 2) have stated: "... despite the increasingly global nature of science communication, research activities in the field are shaped by national

---

Box 10.2: Sources for science communication research

Guenther & Joubert (2017) systematically analyzed science communication as a field and confirmed a recent increase in science communication research articles. They conclude that the vast majority of the authors published only once in the main journals in the field and that most articles were written by one or two authors. A majority of the authors, in particular in the past, were male while recently female authors have become more dominant. They found trends toward more multi-authored papers, as well as more international collaboration. However, they also concluded that more research outputs from Africa, Asia, and the Middle East are needed, which is what Xu, Huang & Wu (2015) also concluded when mapping science communication scholarship in China.

Many academic studies can be found in three dedicated journals: *Public Understanding of Science, Science Communication*, and *JCOM*, the *Journal of Science Communication* (Guenther & Joubert, 2017). The last of these — *JCOM* — is an open-access journal that often includes practically oriented research articles. In addition to these journals, outcomes of science communication studies are also being published in other journals, such as the *International Journal of Science Education, part B Communication & Engagement; Social Studies of Science; New Genetics & Society;* the *Journal of Responsible Innovation;* and *Risk Perception.* In addition, top scientific journals such as *Science* and *Nature* sometimes publish science communication research. In 2018, the English language journal *Cultures of Science* was launched in China to foster science communication research in Asia.

In addition to academic articles, other sources for science communication findings can be found in gray literature, such as reports or popular articles. Reports are often written as part of projects commissioned by different organizations such as governmental parties, research agencies, national research foundations, NGOs, or others. These will sometimes be available online. Outcomes of science communication research are sometimes translated into popular articles for the mainstream media or for online publications like *The Conversation* (*theconversation.com*), which are then easier to understand and accessible for a larger audience.

Open access on the rise
Worldwide, the trend is open-access publications. Open access makes research findings more widely accessible. However, as researchers at universities are often assessed by the quality of the journal that publishes

---

Box 10.2: (*Continued*)

their research, this often discourages them from publishing in open-access journals that do not have the status of highly established journals yet to accept open access.

The number of journals and online forums emerging from African, Asian, and South American regions is slowly rising, emphasizing a trend to balance in a very small way the dominance of western academic publishing models.

---

priorities and cultural contexts and research outputs therefore display regional differences."

Better understanding of the drivers and impediments to research and the factors that shape its directions helps academics and practitioners to undertake better research. Several projects have looked at the differences between practice and research to better understand impediments to them working more closely together. One of the key differences that has been found is that practice moves quickly and is in a near-constant state of innovation, while academic research tends to move slowly and its learnings are incremental (Featherstone *et al.*, 2013). Other differences pointed out by Featherstone *et al.* (2013, pp. 11–13) include the following:

- practitioners can find it hard to easily access collective academic knowledge;
- since many science communicators are originally trained as scientists, they can find the academic language of humanities-based or social-sciences-based researchers 'opaque and challenging' to them;
- academics who study science communication tend to develop their insights through traditional academic means, which is generally in progressively accumulative steps, and this may not even be intended to influence the next thing they work on or have a practical application in sight
- science communication practitioners want academics to prove long-term impacts, while academics tend to see practitioners as research subjects, rather than as collaborators.

These differences can be important to understand as they tend to make collaboration between the two groups harder. To bring the two groups closer together, for instance, the above-mentioned NAS report (2017) focuses on points of important intersection, such as studies on cognitive biases and how they impact practice. It is worth explaining this field of research and how it has focused researchers and practitioners on working more closely together.

Cognitive bias research is fairly new for science communication, but has had a large impact on it. It is the study of how people think, trying to better understand what biases are, and when people are and are not receptive to science messages. A major challenge for science communication has been the finding that people tend to make decisions based primarily on their social, political, moral, and religious values, rather than on information. This also helps explain why members of the public can consistently score low on knowledge of science, but still hold very strong opinions on these subjects (Nisbet & Markowitz, 2016).

Ongoing research into the way humans process information has been looking at, for instance, why some people believe that windmills make people sick even without any evidence of negative health impacts. Or why some people do not trust genetically modified foods despite any evidence of harm to human health. Importantly, this research can also be used to develop strategies that help to counter some of the strongly held beliefs that can impede the development of some technologies, which is particularly evident when there has been no, or poor, consultation with the community about their development which is why practitioners have been very engaged by it.

Climate change research has also been a significant area where research and practice have found common ground, often driven by the political imperative to defend the science and acceptance of climate change impacts. However, if academics have learned anything from all the years of research that has been conducted into trying to better communicate climate change impacts and achieve behavior changes as a result, it is that it is very complex and the more that is learned the more there still seems to be learned. Despite that, studies show that a considerable amount of contemporary climate change communications are quite sophisticated, due to the quality of science communication research that has been conducted in this field.

10.6 Research Questions Left to Ask

If researchers and practitioners are to work better in each other's interests, it is useful to consider the types of research questions that will benefit both of them. Nisbet & Markowitz (2016) looked at the key areas where research and practice were converging and cited research into cognitive biases particularly, describing it as some of the most impactful research occurring, because it not only identifies but also suggests possible effective strategies for overcoming biased information preferences for controversial science.

Contentious science and technologies like GM foods, nanotechnology, fracking, and vaccination have been the subject of many research projects that collectively improve the way that science communication is now done. However, as has been pointed out by Nisbet & Markowitz (2016), much of the research in these fields has been good at describing the social context within which science communication takes place, but often lacks the type of direct recommendations that practitioners prefer. So clearly, while some progress is being made, more work needs to be done on how to use science communication research to improve science communication practice.

The National Academies of Sciences in the US have run a series of meetings thereby bringing together scholars and practitioners to specifically address the gaps between them. The report (NAS, 2017) recommends a research agenda for more effective science communication and proposes the following research areas where more work needs to be done, which will help science communication practitioners and researchers:

- rather than engaging in communication first and evaluating later, ensure the evaluation best suits the communication,
- a better understanding of the impact of social media — both positive and negative,
- a better understanding of which types of communication approaches are best for different audiences, to achieve particular goals,
- understanding how science communicators approach debunking misinformation in a way that is consistent with the weight of scientific evidence, and
- a better understanding of framing, by which information is presented in a certain light to influence how people think, believe, or act.

Other specific areas that have been identified as needing more research to better understand how to undertake science communication activities are as follows:

Public Diversity

It is important to better understand public diversity, which should include those who are not interested in science. Too many science engagements are conducted with people who self-select to participate, which biases the data toward those who are keen to be engaged. Indeed science communicators know very little about those who do not really have an interest in science, who in turn know very little about science communication activities (Cormick, 2014). And only through better understanding the complex and multiple expectations and motivations that different publics have to engaging with science will it be easier to meet those expectations (Mohr, Raman & Gibbs 2013).

Evaluation

Several researchers have criticized the quality of evaluation that is undertaken by practitioners, with Featherstone *et al.* (2013) describing it as often being undertaken for short-term accountability, simply to satisfy a funders' needs for a specific activity. On the contrary, findings obtained from academic evaluations do not always suit practitioners, as the research may be seeking longer term findings. Practitioners will benefit enormously though from focusing on the research being done into effective evaluation, looking more at what needs to be measured rather than what is easy to measure. For instance, it is easy to measure the number of people who attend an event, but that is not the same as measuring how people were impacted by an event or had their awareness or understanding of a topic raised. One of the challenges for effective evaluation is finding a tool that will work across the variety of science communication activities that can exist, which may encompass public talks, debates, exhibitions, publications, science theater, television documentaries, and citizen-led projects. Nevertheless, there are enough models of effective evaluation undertaken by researchers that science communications practitioners should be able to choose the methods that best work for the activities they are undertaking.

Social Media

The rapidly changing nature of social media, such as Facebook, presents quite a challenge for both researchers and practitioners to keep up with and also demands ongoing research to feed into best practice. One of the main challenges is that research into social media can become outdated very quickly. For science communicators trying to build greater public engagement with quality sources of news and information about science, understanding the role of, for example, Facebook friends in cultivating attention to news stories or growing the audience for quality sources of information is very important (Nisbet & Markowitz, 2016).

Aligning Information with Audiences

Scientists and practitioners are not always good at knowing what an audience wants or needs to help them understand a complex topic. This is because their expertise and experience may make them too close to an issue and unable to see how a member of the public might view it. Therefore, research into better understanding audiences helps a practitioner in developing the information the audience most wants and needs.

Narratives and Storytelling

While a lot of researchers agree that stories can be very powerful ways to communicate science, few have studied stories as a way to create better persuasive narratives. Dahlstrom (2014), for instance, has described the way narratives work by using a particular voice to set up a conflict, unresolved question, or tension around a science-related debate. A better understanding of precisely how this works in particular situations or settings would be of significant benefit to both researchers as well as practitioners.

The Impact of Entertainment Media

After formal education, entertainment and news media have become the dominant sources of information about science and technology for most people (Eveland & Cooper, 2013). More research is needed on how entertainment creates or reinforces attitudes to science, particularly when much of it is done through a fictitious narrative frame.

Countering Misinformation or False Beliefs

Certainly, the politicization of science is nothing new, but a renewed emphasis on 'fake news' and misinformation has emerged in recent years, driven by global politics and a better understanding of cognitive biases. Nisbet & Markowitz (2016) stated as follows:

> This renewed interest in understanding the root causes of misinformation, politicization, and false beliefs is undoubtedly driven at least in part by well-documented efforts to undermine the credibility of science in numerous domains, including public health, climate change, and environmental conservation (p. 37).

Other research (Lewandowsky *et al.*, 2012) has already shown that well-intentioned but ill-conceived, intuitive efforts to debunk misinformation often end up with the unintended effect of reinforcing false beliefs (Nyhan & Reifler, 2010). This demonstrates the necessity of basing science communication practice on good research. See Box 10.3 for details.

Box 10.3: Communicating science over pseudoscience: A case study from South Africa

by Marina Joubert and Nokwanda Makunga.

Being able to apply science communication research theories to best practice is important for most forms of science — but sometimes, it can be a matter of life and death, particularly when trying to combat popular, but unscientific, health claims.

Box 10.3: *(Continued)*

For instance, many people around the world trust practices that may appear to be scientific on the surface, but lack credible evidence. When dubious claims are disguised as reliable knowledge, often with a veneer of science, it fits the broad definition of 'pseudoscience'. Age-old examples include astrology, homeopathy, reiki, and detox diets, which have not been able to show any credible scientific-based benefits. From time to time, new bogus products such as 'power balance bracelets' and 'ionic footbaths' appear on the market. Many people also discard reason and logic when it comes to faith healing, psychics, mediums, clairvoyants, and superstitions, clearly showing that pseudoscientific beliefs and practices are widespread in both the developed and developing world.

So, what is different from a South African perspective?

For centuries, South Africans have relied on indigenous plants for medicines. Knowledge about the healing properties of plants is passed down from one generation to the next and preserved in folklore. Traditional healers are the custodians of knowledge of medicinal plants, but their complex and multi-layered practices also include a wide variety of animal parts (see Johnson, 2013), often used for pseudo-therapeutic effects such as making someone stronger, boosting luck, or warding off bad spirits (Williams & Whiting, 2016).

In many rural and urban communities in South Africa, consultations with traditional healers are deeply embedded in the traditional way of life and therefore historically and culturally significant. The wares of the traditional healers are often sold on sidewalks and informal markets, but some healers run modern health practices and offer their services via web sites and social media. Their services and qualifications are mostly unregulated (see Street & Rautenbach, 2016) (Figure 10.2).

In cases where the beneficial properties of local plants have been validated scientifically, these plants are used for the development of new products (Makunga, Philander & Smith, 2008). Unfortunately, uncertainties about the boundaries between evidence-based science, indigenous knowledge systems, and traditional beliefs create gaps for pseudoscience to flourish.

While some traditional healers are willing to cooperate with scientists toward proving the efficacy of their treatments (Ramchundar & Nlooto, 2017), this is not the case for bogus healers. It can also be difficult

---

Box 10.3: (*Continued*)

for scientists to challenge some of the pseudoscientific practices, because they are deeply ingrained in local cultures. Consequently, people continue to be misled by 'quacks' posing as authentic herbalists and continue to pay for dubious remedies in return for promised solutions to problems related to work, money, and love.

Figure 10.2: Plant materials on sale at a herbal market in South Africa (Photo credit: Dr Lisa Philander).

Reliance on pseudoscience is, of course, not limited to poorer people living in rural areas. One of the most devastating examples of how dependence on pseudoscience can lead to loss of life resulted from AIDS denialism by South Africa's second democratically elected president, Thabo Mbeki, which peaked around 2000. Mbeki questioned the science of HIV/AIDS and the existence of a virus that caused this disease (Schneider & Fassin, 2002; Mbali, 2004). By his side, health minister Dr. Mantombazana Tshabalala-Msimang promoted a variety of untested therapies via the media, insisting on a concoction of garlic, beetroot, and lemon as a weapon against HIV/AIDS (Nattrass, 2007). It is estimated that the delay

Box 10.3: (*Continued*)

in implementing the antiretroviral program in South Africa led to the deaths of more than 330,000 people, while about 35,000 HIV-positive babies were born during this period (Chigwedere *et al.*, 2008).

Even today, with antiretroviral treatment available in the public health sector, pseudoscientific beliefs and health scams continue to thrive around potential cures for HIV/AIDS. In the region of KwaZulu-Natal, a mixture called 'uBhejane' has been widely promoted as a curative tonic for HIV/AIDS (Cullinan, 2006). Its popularity is perceived to be linked to deep-rooted suspicions of the so-called 'white' medicine (Batts, 2006). 'People are desperate, and they want to believe in traditional medicine in a way that is probably not true in other settings,' says South African epidemiologist and infectious diseases specialist Professor Salim Abdool-Karim (Joubert, 2018, p. 245). 'We found that "uBhejane" came from an out-of-work truck driver who filled bottles with coloured water and sold it as HIV medication.'

Furthermore, the persistent myth that someone who is HIV positive can be cleansed or even cured by having sex with a virgin leads to heartbreaking consequences in the form of the rape of infant girls (Pitcher & Bowley, 2002). These tragic occurrences may increase when people do not have access to modern medicine (Leclerc-Madlala, 2002).

While some herbalists are able to help their clients with their knowledge of the medicinal properties of plants, a mistaken belief in their powers may have deadly consequences. During the Marikana massacre on August 16, 2012, 34 striking mineworkers were shot dead by police. Reportedly, they believed that the *muthi* — a plant- or animal-based substance prepared by herbalists that is believed to heal, cleanse, strengthen, or protect — they took would make them invincible in the face of bullets (Nyundu & Naidoo, 2016).

Because of its close integration with traditional health care and folklore, people rely on pseudoscientific practices in South Africa in a number of ways that are cause for concern. While scientists may perceive a collective duty to combat this kind of misinformation, evidence shows that just labelling the practices as harmful or unscientific does little to change people's attitudes. Added to that is a need to better understand the complex challenges around being sensitive to local sociocultural dynamics. This means, in effect, that science communication solutions need to be based on research undertaken in the impacted community, and research theories

(*Continued*)

Box 10.3: (*Continued*)

developed in other countries can rarely be imported without adapting them to a local context.

References

Batts, S. (2006). *Science blogs. Traditional African potion as AIDS "Cure"?* (No, just Woo). Retrieved September 15, 2018 from http://scienceblogs.com/retrospectacle/2006/08/23/traditional-african-potion-as-1/.

Chigwedere, P., Seage, G. R., Gruskin, S., Lee, T. H., & Essex, M. (2008). Estimating the lost benefits of antiretroviral drug use in South Africa. *Journal of Acquired Immune Deficiency Syndromes, 49*(4), 410–415.

Cullinan, K. (2006). Health official promote untested uBhejane. *Health-e News.* Retrieved April 5, 2019 from https://www.health-e.org.za/2006/03/22/health-officials-promote-untested-ubhejane/.

Johnson, D. (2013). Monkey paws at Durbans Muthi market. *Africa Geographic.* Retrieved April 5, 2019 from https://africageographic.com/blog/monkey-paws-at-durbans-muthi-market/.

Joubert, M. (2018). *Factors influencing the public communication behaviour of publicly visible scientists in South Africa.* PhD thesis. Stellenbosch: Stellenbosch University South Africa.

Lecrerc-Madlala, S. (2002). On the virgin cleansing myth: Gendered bodies, AIDS and ethnomedicine. *African Journal of AIDS Research, 1*(2), 87–95.

Makunga, N. P., Philander, L. E., & Smith, M. (2008). Current perspectives on an emerging formal natural products sector in South Africa. *Journal of Ethnopharmacology, 119*(3), 365–375.

Mbali, M. (2004). AIDS Discourses and the South African State: Government denialism and post-apartheid AIDS policy-making. *Transformation: Critical Perspective on Southern Africa, 54*(1), 104–122.

Nattrass, N. (2007). *Mortal Combat: AIDS Denialism and the Struggle for Antiretrovirals in South Africa.* Scottsville: University of KwaZulu-Natal Press.

Nyundu, T., & Naidoo, K. (2016). Traditional healers, their services and the ambivalence of South African youth. *Commonwealth Youth and Development, 14*(1), 144–155.

Pitcher, G., & Bowley, D. (2002). Infant rape in South Africa. *The Lancet, 359*(9303), 274–275.

Ramchundar, N., & Nlooto, M. (2017). Willingness of African traditional healers to collaborate with researchers to further develop their

Box 10.3: *(Continued)*

traditional medicines in Kwazulu-Natal Province of South Africa. *PULA: Botswana Journal of African Studies, 31*(1), 163–179.

Schneider, H., & Fassin, D. (2002). Denial and defiance: A socio-political analysis of Aids in South Africa. *AIDS, 16*(Suppl. 4), S45–S51.

Street, R., & Rautenbach, C. (2016). South Africa wants to regulate traditional healers but its not easy. *The conversation.* Retrieved April 5, 2019 from https://theconversation.com/south-africa-wants-to-regulate-traditional-healers-but-its-not-easy-53122.

Williams, V. L., & Whiting, M. J. (2016). A picture of health? Animal use and the Faraday traditional medicine market, South Africa. *Journal of Ethnopharmacology, 179,* 265–273.

Ethics

Different from research ethics, undertaking research into ethical issues related to science communication is a growing field of attention for researchers as well. A wide range of topics are being looked at, including the ethics of hyping science stories to get media coverage, deciding which science stories to promote and which not to and how to use framing and narratives in a way that is not biasing the data. Some researchers have called for science communicators to develop a code of ethics, but it has been pointed out that since science communication as a discipline has come from so many different areas, none of the existing ethics models from journalism, social sciences, or other sciences are fully compatible with each other. Much more research and discussion is likely to take place in this field before any clarity is seen on the type of ethics or ethics models which will work best for science communication.

In addition, in terms of research into science communication a few pertinent questions are not being asked enough. They include issues with reproducibility, where one-off studies are held up as gold standards, and there is a bias toward experiments that find a result over those that do not, as these studies do not have much more trouble getting published (see Box 10.4).

Box 10.4:   Research questions that are not being asked enough

More questions should perhaps have been asked about reproducibility issues in science communication studies when the Reproducibility Project undertook a major study of psychology research in 2015. The project involved over 270 psychology researchers trying to replicate the findings of 100 key research studies from 2008 (Estimating the reproducibility of psychological science, 2015).

The project found that only about a third of them could be replicated. This sent researchers in other fields to look at replicating other key studies, and similarly, it was found that many social science studies did not have a high rate of being able to be replicated.

In the science communication space, a similar outcome was found for a highly cited paper on the cognitive biases. Wood & Porter (2019) replicated a paper by Nyhan & Reifler (2010) that found that if people's beliefs were shown to be incorrect, they would actually become more entrenched in those beliefs. This became known as the 'backfire' principle. Wood and Porter increased the sample size from the original study from 327 undergraduates to 10,100 subjects, and then exposed them to the same misleading claims made by US political figures, as well as the corrections to them, that were used in the Nyhan & Reifler (2010) study. They found that for only one of the 36 topics being corrected, there was any backfire. This was a statement around the existence of weapons of mass destruction.

Of course, different studies held at different times with different audiences will achieve different results, to some extent. But given the fact there is little incentive to reproduce most social science or science communication studies, researchers and practitioners might need to closely consider how well single studies can be projected across times and cultures and be more cautious in widely extrapolating the findings of any single study.

10.7   Conclusion: How to Work Toward each Other's Interest?

This chapter provided some insights into science communication research. The key points from this chapter are that the science communication community can be divided into roughly two parts: those doing practice and those engaged in research. For representatives of both groups, setting up a research project will follow the same steps from formulating the research aim and research questions, to writing up the results. Key methods for conducting research include qualitative and quantitative data collection, with the main methods being surveys, interviews, focus groups, and case studies. Nowadays, more attention for research ethics is part of every study.

Finally, in outlining the way that research into science communication is done, the divide between research and practice was looked at, and it was argued that both groups would increasingly benefit if research was more focused on assisting practice (and practice more focused on helping research). Of course, not all research needs to benefit practice and not all practice needs to benefit research, but closer work between the two will ultimately benefit both.

References

Bauer, M. W. (2009). The evolution of public understanding of science — Discourse and comparative evidence. *Science, Technology & Society, 14*(2), 221–240.

Bauer, M. W., Allum, N., & Miller, S. (2007). What can we learn from 25 years of PUS survey research? Liberating and expanding the agenda. *Public Understanding of Science, 16*(1), 79–95.

Brewer, J., & Hunter, A. (2006). *Foundations of Multimethod Research. Synthesizing Styles.* Thousand Oaks, CA: Sage.

Cormick, C. (2014). *Community Attitudes towards Science and Technology in Australia.* Canberra: CSIRO.

Dahlstrom, M. F. (2014). Using narratives and storytelling to communicate science with nonexpert audiences. *Proceedings of the National Academy of Sciences of the United States of America, 111*(Supplement 4), 13614–13620. doi:10.1073/pnas.1320645111.

Dijkstra, A. M. (2008). *Of publics and Science. How Publics Engage with Biotechnology and Genomics.* Thesis, Enschede: University of Twente.

Egan, A. F., Jones, S. B., Luloff, A. E., & Finley, J. C. (1995). The value of using multiple methods: An illustration using survey, focus group, and delphi technique. *Society and Natural Resources, 9*, 457–465.

Eveland, W. P., & Cooper, K. E. (2013). An integrated model of communication influence on beliefs. *Proceedings of the National Academy of Sciences of the United States of America, 110*(Supplement 3), 14088. doi:10.1073/pnas.1212742110.

Estimating the reproducibility of psychological science. (2015). *Science, 349*(6251), aac4716. doi:10.1126/science.aac4716.

Featherstone, H., Manners, P., Nerlich, B., and James, H. (2013). Science communication — bridging theory and practice. In *Science Communication: State of the Nation 2013*, Essays inspired by the annual Science Communication Conference the Royal Institution. London: British Science Association. Retrieved March 28, 2019 from http://ow.ly/tduWr

Greene, J., Benjamin, L., & Goodyear, L. (2001). The merits of mixing methods in evaluation. *Evaluation, 7*(1), 25–44. doi:10.1177/13563890122209504.

Guenther, L., & Joubert, M. (2017). Science communication as a field of research: Identifying trends, challenges and gaps by analysing research papers. *Journal of Science Communication, 16*(2), 1–19.

Jensen, E. A., & Laurie, C. (2016). *Doing Real Research. A Practical Guide to Social Research*. London: SAGE.

Lewandowsky, S., Ecker, U. K., Seifert, C. M., Schwarz, N., & Cook, J. (2012). Misinformation and its correction continued influence and successful debiasing. *Psychological Science in the Public Interest, 13*(3), 106–131.

Mohr, A., Raman, S., & Gibbs, B. (2013). *Which publics? When? Exploring the policy potential of involving different publics in dialogue around science and technology*. Nottingham: Sciencewise. Retrieved March 26, 2019 from https://webarchive.nationalarchives.gov.uk/20170110114316tf_/http://www.sciencewise-erc.org.uk/cms/which-publics-when/.

Mulder, H. A. J., Longnecker, N., & Davis, L. S. (2008). The state of science communication programs at universities around the world. *Science Communication, 30*(2), 277–287. doi:10.1177/1075547008324878.

National Academies of Sciences, Engineering, and Medicine (NAS) (2017). *Communicating Science Effectively. A Research Agenda*. National Academy of Sciences: Washington, DC. Retrieved March 26, 2019 from https://www.nap.edu/catalog/23674/communicating-science-effectively-a-research-agenda.

Nisbet, M. C., & Markowitz, E. (2016). Science Communication Research: Bridging Theory and Practice. Commissioned Synthesis and Annotated Bibliography in Support of the Alan Leshner Leadership Institute of the American Association for the Advancement of Science, Washington, DC: AAAS. Retrieved March 26, 2019 from https://www.aaas.org/sites/default/files/content_files/NisbetMarkowitz_SciCommAnnotatedBibliography_Final.pdf.

Nisbet, M. C., & Scheufele, D. A. (2009). What's next for science communication? Promising directions and lingering distractions. *American Journal of Botany, 96*(10), 1767–1778.

Nyhan, B., & Reifler, J. (2010). When corrections fail: The persistence of political misperceptions. *Political Behavior, 32*(2), 303–330.

Petticrew, M., & Roberts, H. (2006). *Systematic Reviews in the Social Sciences.* Malden, MA: Blackwell Publishing.

Schuurbiers, D., & Fisher, E. (2009). Lab-scale intervention: Science & society series on convergence research. *EMBO Reports, 10*(5), 424–427. doi:10.1038/embor.2009.80.

Sturgis, P., & Allum, N. (2004). Science in Society: Re-evaluating the deficit model of public attitudes. *Public Understanding of Science, 13*(1), 55–74. doi:10.1177/0963662504042690.

Trench, B., & Bucchi, M. (2015). *Science communication research over 50 years: patterns and trends.* Paper presented at the Science and You. Retrieved March 26, 2019 from https://www.academia.edu/14286143/Science_communication_research_over_50_years_patterns_and_trends.

Von Grote, C., & Dierkes, M. (2000). Public understanding of science and technology: State of the art and consequences for future research. In M. Dierkes and C. von Grote (eds.). *Between Understanding and Trust. The Public, Science and Technology.* Amsterdam: Harwood, pp. 341–361.

Wilkinson, C., & Weitkamp, E. (2016). *Creative Research Communication. Theory and Practice.* Manchester: Manchester University Press.

Wood, T., & Porter, E. (2019). The elusive backfire effect: Mass attitudes' steadfast factual adherence. *Political Behavior, 41*(1), 135–163. doi:10.1007/s11109-018-9443-y.

Wynne, B. (1995). Public Understanding of Science. In S. Jasanoff, G. E. Markle, J. C. Petersen, & T. Pinch (eds.). *Handbook of Science and Technology Studies.* London / New Delhi: Sage, pp. 361–388.

Xu, L., Huang, B., & Wu, G. (2015). Mapping science communication scholarship in China: Content analysis on breadth, depth and agenda of published research. *Public Understanding of Science, 24*(8), 897–912. doi: 10.1177/0963662515600966.

Index